career

CONNECTIONS

Oxyfuel Cutting and Shielded Metal Arc Welding

Career Connections: Oxyfuel Cutting and Shielded Metal Arc Welding CCWELDM
July 2018
ISBN: 978-0-692-92295-8
123456789 17 98765432

Send all inquiries to:
Carpenters International Training Fund
212 Carpenters Union Way, Las Vegas, NV 89119-4218
www.carpenters.org

Associate Membership ACTE

preface

Introducing students to new trades and helping them to improve their existing skills is a top priority for the Carpenters International Training Fund (CITF). We consider technical and employability skills, lifelong training, and productivity to be the keys to career success. We offer skill training to high school students in carpentry and millwrighting and skills relevent to these crafts, such as welding.

Excellent training involves transferring skill and knowledge from highly skilled and experienced teachers to students whose interest has brought them to a basic welding class. Because the authors of these materials have both classroom and field experience, these materials help provide a solid foundation on which to build a future in the construction industry.

A Cautionary Note

This manual is intended as a supplement to actual hands-on instruction and is designed to teach one or more of the acceptable and recognized methods of performing specific tasks. It is not meant to be, nor should it be considered, an absolute or complete presentation of the procedures and safety measures that relate to these tasks.

Work processes and government safety regulations can and do change, and it is the employer's responsibility to provide workers with the most recent technical and safety information involving these processes. The guidelines and instructions presented here are not meant to supersede manufacturers' instructions or contractors' jobsite procedures, nor are they meant to replace any current local, state, provincial, or federal safety rules or regulations.

It is essential that you always follow all current local, state, provincial, or federal safety rules, regulations, and guidelines whenever you perform any of these tasks.

No statements made in this manual should give the impression that the Carpenters International Training Fund or the United Brotherhood of Carpenters and Joiners of America, their affiliates, representatives, or employees have assumed any part of the employer's legal responsibility to provide a "safe and healthful workplace," as mandated by the Occupational Safety and Health Act of 1970.

acknowledgments

The preparation of this training material would not have been possible without the technical contributions and cooperative efforts of a wide range of individuals and groups associated with the United Brotherhood of Carpenters.

Let us start with the members. Without their commitment to excellence, their willingness to seek training and their continued contributions and support, this program would be meaningless. It is your contributions that make the program possible, and it is your interests we seek to serve.

We would like to acknowledge the teachers and JATCs throughout the U.S. and Canada who helped develop and review this curriculum. We also recognize the contributions of the members of the Carpenters International Training Advisory Group. Finally, we recognize the ongoing efforts of the staff at the Carpenters International Training Fund, which are vital to keeping the project on track and continue to be key to a successful implementation of the curriculum.

In particular, we owe a special thanks to the following individuals who served as subject matter experts for this curriculum and the Training Programs they represent, which generously allowed them to participate in the project.

Harry Bottino
Carpenters Joint Apprenticeship Committee of Philadelphia

Mary Cook
Southern Illinois Carpenters Joint Apprenticeship Program

G. William Hands
Carpenters Training Trust of Western Washington

Douglas Lockhart
Southern Nevada Carpenters and Millwrights Apprenticeship and Journeyman Training Trust

John Marconi
Carpenters Joint Apprenticeship Committee of Philadelphia

John Vincent
Southern Nevada Carpenters and Millwrights Apprenticeship and Journeyman Training Trust

contents

Oxyfuel

SMAW

Weld Symbols

introduction

The Welder: Job Duties and Employability Skills

Overview

Welding is a skill that is a subset of carpentry and millwrighting. Both Millwrights and Carpenters have a variety of jobs that require welding to achieve quality results. The job duties of welders are among the most critical of all the construction trades in building and maintaining industrial systems. Their job duties require a technical skill set necessary for installing and maintaining structures. Along with these technical skills, many job duties also require physical skills and other abilities. The most successful welders also possess skills that employers have identified as those that allow workers to excel in all types of job conditions. These are called employability skills and will be discussed later in this introduction.

Job Duties

Because of the wide range of industries that require welding skills, job duties are especially varied and extensive. These duties involve constructing, installing, maintaining, and upgrading structures according to a prescribed set of plans, prints, and other drawings. In addition, job duties involve the following:

- dismantling and assembling structures
- selecting rigging equipment according to the weight and size of the material to be moved
- moving and positioning pieces of structures
- layout and fabrication
- reading prints and schematic drawings
- working on structural foundations of buildings, bridges, and power plants

Some duties require teamwork with fellow welders as well as other workers in crafts unrelated to welding. At times, job duties also require the ability to adapt to difficult working conditions. Some jobsites may be very noisy, such as paper mills, or extremely hot, such as steel mills. Whatever the conditions on the jobsite, work must be done in a safe and efficient manner. Safety training is a major part of all construction work.

Physical Requirements

Along with the technical skills used for installing and repairing structures, welders must also possess several physical skills and abilities, including hand-eye coordination, balance, and agility. For example, a task may involve handling tools and other equipment while maintaining body balance when working from scaffolding or other high structures. Many jobs require stooping, laying, bending, squatting, kneeling, crouching, or crawling for prolonged periods of time. Tools, equipment, and materials may also have to be lifted and carried. Knowledge of ergonomics, which is the study of how the work, the worker, and the workplace fit together, is important to a welder. Using proper body positioning and ergonomically designed tools is a proactive approach to prevent injuries and is one of the most effective ways to remain injury free.

Employability Skills

Unfortunately, not everyone who goes to work as a welder achieves a high level of success or job satisfaction. What typically makes a person succeed or not succeed is based not only on the technical skills needed but also on the skills needed to be employable. Employers have identified a set of skills, called employability skills, which are aptitudes and traits other than technical and physical abilities that employers expect workers to possess. A great deal of research has gone into what makes a person

employable and identifying the skills necessary to succeed in the workplace. Most research indicates that to be successful a person must possess the following skills, among others:

- ability to work with others
- ability to understand and use information
- ability to plan time and resources
- ability to use different technologies
- ability to understand complex interrelationships

Work With Others

The most fundamental skill necessary to succeed is the ability to get along with others. This includes fellow employees and supervisors, as well as clients and other people on the jobsite. Being able to work with others is not as simple as getting along; it also means understanding the needs of the workplace. This can be as simple as arriving at work on time so the job is not delayed and members of the work crews do not have to be reassigned to cover for an absent crew member.

Teamwork is a major part of working at a jobsite. Seldom does one person accomplish anything alone. For example, when a welder is making a weld, another worker, called a fire watch, is standing by to protect the welder from catching on fire and to protect the surrounding equipment. The entire crew relies on teamwork for its safety.

Understand and Use Information

An essential part of a welder's job is to gather and accurately interpret information at a jobsite. This information may be in the form of prints or shop drawings, or it may be verbal or written instructions. It may consist of mathematical equations and dimensions, or it may be a list of materials and hardware necessary to construct a platform or repair a machine. It might be job-specific or site-specific, and it may include safety information that is critical to act on. Being able to understand and use information requires asking questions when meanings are unclear.

Plan Time and Resources

In the workplace, it is not enough just to perform multiple tasks; those tasks must be performed accurately and according to the schedule. Concentrating on the task itself and not the scope of the entire job may mean that the task must be repeated because it was done out of order. The craft of welding involves the use of different tools and materials, many of which can be very expensive. Cutting a beam the wrong dimension can result in an unnecessary expense for the job and wasted productivity time.

Use Different Technologies

Computers are a part of modern society. It is hard to find any job that is not directly affected by information technology. Computers and the programs they run are deeply involved in how the workplace operates. Software programs such as Building Information Management (BIM) or Plan Grid® are used to create and read prints. Welders need to be able to use these programs to consult prints electronically. Keeping up with the advances in technology is important to career success.

Understand Complex Interrelationships

Every person has individual needs. The same is true with every organization. An organization operates according to its needs and the needs of its members. To be successful, it is important to understand the needs of the organization; doing so will allow the individual to understand why decisions were made. However, it is not enough to simply understand. Understanding the organization means helping it improve. Suggestions that help improve an existing way of doing things or that develop a new way to do them are always welcome in any organization, especially if those improvements save time and money.

The Foundation for Employability Skills

Just as with a building, employability skills require a solid foundation. Welders must have a preexisting set of skills from which to build employability skills. The skills or qualities that make up the three-part foundation of employability skills include basic skills, thinking skills, and personal qualities.

Basic Skills

Basic skills include skills that have long been taught in schools: reading, writing, and arithmetic. However, they also involve basic communication skills, such as listening and speaking. All of these skills are used extensively in the workplace and are intertwined with every one of the employability skills. The ability to read and write well is essential. Transferring dimensions and taking accurate measurements requires strong math skills. Being able to work with others, as well as gathering and understanding information, requires the ability to speak clearly and to listen well.

Thinking Skills

Thinking skills do not just include problem solving. They also involve being able to make a decision, which is much harder to do without the ability to gather and understand information. One of the most important thinking skills is knowing how to apply new knowledge and skills. This is especially important because the worker is constantly being given new information to process and then is shown the skills that must be developed to complete assigned tasks.

Personal Qualities

Probably the most important of all the employability skills are the personal qualities that an individual possesses. These include such qualities as integrity—strictly following a moral code, self-esteem—respect for oneself, and honesty. They also include friendliness, adaptability, and empathy—understanding another's feelings or situation. All of these qualities are important when working with others. In addition, self-management is an important personal quality that will dictate how well a person plans time and uses resources.

CHAPTER 1

Welding and Cutting Safety

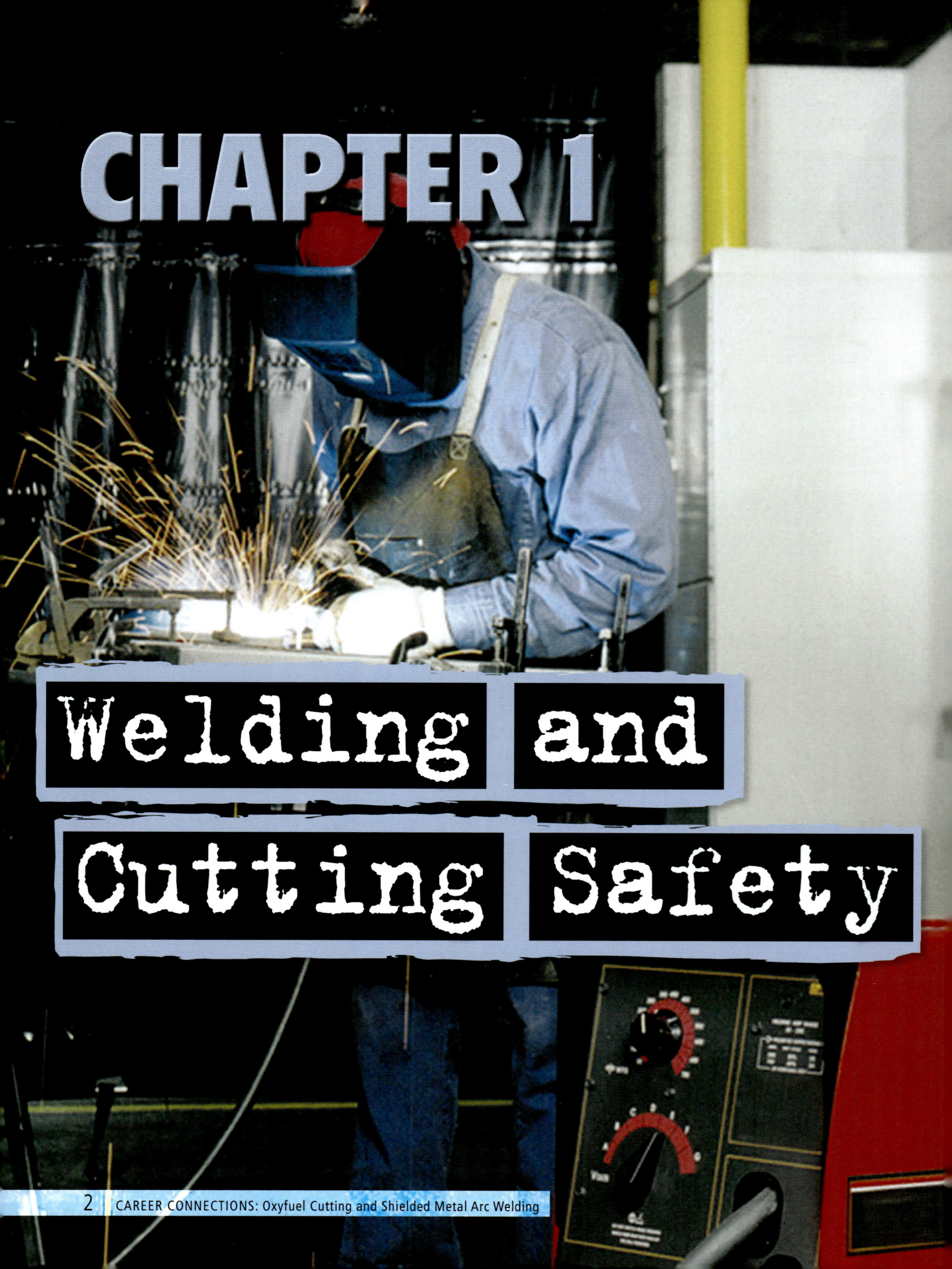

Welding metals together is not a gentle process. The act of welding causes intense light, noise, smoke, and fumes. Welders work in the presence of gasses and high heat. Protective clothing such as jackets, chaps, helmets, shields, and so on, is very important to protect the welder. Knowing how to work safely where welding is occurring will protect not only the welder but also coworkers and property.

Contents

Objectives

- Identify the hazards associated with welding.
- List the proper clothing to wear when welding.
- Demonstrate an understanding of welding safety precautions.
- Describe personal protective equipment and its use.
- State the importance of personal protective equipment.
- List the materials and parts that helmets are made of.
- Describe the different styles of helmets.

Key Terms

bib part of a leather ensemble to protect the upper torso

chaps protective covering to provide extra leg protection against heat and sparks or hot particles

face shield protected covering for the face that can be clear for grinding or shaded for cutting

fire blanket coverings used to protect personnel and equipment

friction lighter device that uses flint and a striking motion to create a spark

fumes vapor, gas, or smoke by-products of welding

hot work any work involving burning, welding, or similar operations capable of initiating fires or explosions

noise unpleasant or loud sounds; can damage hearing

non-ionizing radiation visible, ultraviolet, or infrared radiation; heat or light transfer from welding work

personal protective equipment (PPE) equipment, such as goggles, helmets, gloves, and so on, specifically designed to protect exposed or vulnerable body parts when welding

shade rating assessment of how dark a lens is; measured on a numeric scale—the smaller the number the lighter the shade

sleeves can be worn over existing clothing to protect the arms

spats designed to protect the ankle area from exposure to welding particles

visor shaded or clear viewing window in a welding helmet

welding helmet specially made plastic or leather shield to protect the face and eyes from welding by-products

Figure 1
Welder wearing proper clothing

1 Safety Hazards

Joining metal together takes a great deal of persuasion: heat, pressure, gasses, and electricity are all involved in the process. Any one of these processes can be hazardous. A great deal of care needs to be taken so that people and property are not harmed.

The following hazards and ways to handle them will be discussed: hot work, hot sparks, radiation, metal fumes and smoke, and noise. Each one of these hazards should be taken seriously by the welder. Safety clothing and its functions will also be discussed. The American Welding Society (AWS) has put together a series of Safety and Health Fact Sheets that address welding hazards and how to prevent accidents. These Fact Sheets were used to provide the material below.

Fire Prevention and Protection

Any work involving burning, welding, or similar operations capable of initiating fires or explosions is considered **hot work**. Welding by its very nature involves chemical reactions brought on by heat. This heat can cause nearby items to catch fire or burn someone working on or near welding activity. A welder must always be aware of potential fire hazards.

Hot sparks result from welding and cutting activities. These sparks can fly in all directions and can be a hazard to the welder, those standing near the work, and to objects nearby. In fact, flying sparks are the main causes of fires during welding and cutting activities. Sparks and molten metal can travel up to 35′-0″ horizontally from the work area. In addition, sparks can pass through or become lodged in ceilings, floors, walls, and other small openings where they can cause fire

To prevent fires, welders should be mindful of the following.

- Welding should be done in designated areas when possible. A designated welding area will have signage and be free of combustibles.
- When welding work cannot be moved to a designated area, safe guards such as portable curtains and **fire blankets**—coverings used to protect personnel and equipment—should be used to protect flammable items or personnel from sparks and slag.
- Combustible floors may be wetted down or covered with wet sand or metal. Extra precautions must be taken to help personnel avoid electrocution hazards.
- Care should be taken when using a **friction lighter**, a device that uses flint and a striking motion to create a spark.
- To protect against fires welders should use one or more of the following:
  - fire extinguishing equipment, such as pails of water, buckets of sand, hoses, and portable extinguishers
  - fire watchers—people working with welders to watch for fires caused by flying sparks or slag; fire watchers should be trained on specific procedures in the event of fire, as well as the use of fire prevention and protection

Figure 1 shows a properly dressed welder.

» always be aware of potential fire hazards «

Radiation and Burns

Intense bright flashes of light at a construction site are a sure sign that welding is taking place. That bright light is one of the radiation hazards of welding. There are two types of radiation.

- ionizing radiation
- non-ionizing radiation

Ionizing radiation is present in the dust created during the grinding of thoriated tungsten electrodes for gas tungsten arc welding (GTAW). Ionizing radiation is capable of displacing electrons from atoms, creating charged atoms or ions. Ionizing radiation is *extremely* harmful to all living beings. Excessive exposure to it poses a health risk.

The other kind of radiation, **non-ionizing radiation**,occurs in just about every other kind of welding except resistance, or spot, welding and cold-pressure welding. There are three types of non-ionizing radiation: visible, infrared, and ultraviolet.

The welder can see the visible radiation in the form of light but infrared and ultraviolet cannot be seen. Infrared and ultraviolet light, along with visible light, are present when most welding processes are conducted. Non-ionizing radiation can cause eye damage and skin burns. When arc welding or observing arc welding, it is necessary to observe safe guards, such as proper eye and skin protection, appropriately shaded helmet, gloves, clothing that protects the neck area, and a **face shield**—protected covering for the face that can be clear for grinding or shaded for cutting.

Metal Fumes and Gasses

Metal **fumes** and gasses are by-products of welding and cutting processes. Metal fumes and gasses originate from welding consumables such as coated electrodes, the base metal, or any coatings on the base metal. The amount and makeup of these fumes depends on several factors such as composition of filler and base material and the type of welding process being used. The supervisor must be asked for specific safety-related issues or consult a Safety Data Sheet (SDS), if applicable, before starting.

Care must be taken to avoid contact with these by-products as they pose a health risk. Several things can be done to avoid exposure to fumes and gasses.

- Position the head or the work to avoid the fumes and gasses
- Make sure working area is well ventilated. There are several types of ventilation:
  - Natural—a breeze or natural air movement through space
  - Mechanical—a device such as a fan to move the air
  - Local exhaust—a type of mechanical ventilation which uses a fixed or moveable exhaust hood placed close to the weld which captures the weld fumes and gasses; local exhaust is the preferred method because it resembles a vacuum hose that can be pulled close to the weld piece and can be exhausted away from the welder or filtered and recirculated

safety tip

If safety questions arise about a material being worked with, consult its Safety Data Sheet (SDS).

safety tip

Special care must be taken with stainless steel base and filler metals. Stainless materials contain hexavalent chromium. This is a known carcinogen.

Noise

Welding and cutting and the equipment used can produce high noise levels. **Noise** is sound that changes frequency, high or low pitch, and intensity, degree of loudness, and is generally unpleasant, uncomfortable, and sometimes painful. Noise levels are measured in *decibels*. Regulations require safety precautions if noise levels reach 85 decibels on an 8-hour time weighted average (TWA) basis.

When exposed to high decibel levels for a long duration, hearing loss and other health hazards can occur. Not only is noise a health concern, it can also be a safety concern: it can drown out sounds the welder needs to hear.

There are several ways to protect hearing and health from noise.

- Use room acoustics to reduce or control noise.
- Wear protective earplugs, which offer some protection.
- Wear earmuffs, which offer more protection.
- Wear both together, which offers the most protection.

Figure 2 shows a common type of earplug.

safety tip

Earplugs and protective earmuffs not only offer hearing protection, they also protect welders from sparks, spatter, and hot slag.

Figure 2
Earplugs provide good noise protection

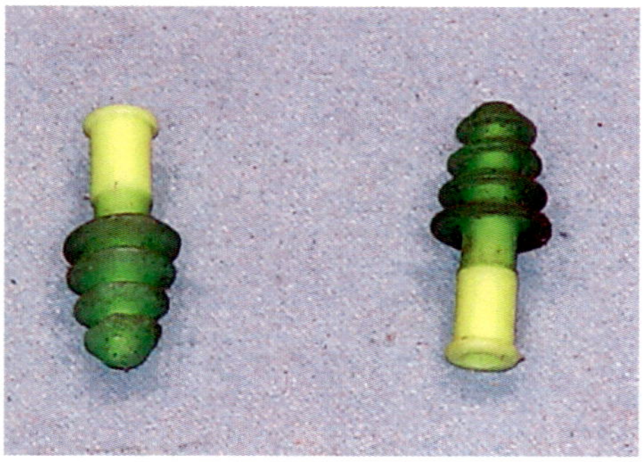

Electrical Hazards

Shielded metal arc welding utilizes electricity to create the welding arc. Any time electricity is used as a tool, there is a danger of shock. Electrodes are electrically charged. To avoid electric shock, do not touch the bare end of the electrode.

Shielded metal arc welding uses voltages ranging from 17 to 45 and approximately from 100 to 500 amps DC or AC. To avoid electric shock, the safety precautions below should be followed.

- Keep all electrical circuits dry.
- Keep all electrical ground connections mechanically tight.
- Only use cables that are designed to carry the amount of voltage that will be used; make sure the cable is in good repair.
- Do not touch live circuits.
- When replacing torch parts, make sure that all power to control circuitry is disconnected.
- Always practice safe welding techniques.

In addition to the rules listed above, always use the ground cable properly. Place the ground cable as near to the work as possible. Placing the ground cable on the workpiece itself is the best practice, if this is possible. Make sure the clamp is firmly attached to the ground surface.

safety tip

To avoid electric shock, do not touch the bare end of the electrode.

safety tip

Always attach the ground cable securely to the workpiece.

Self Check

1. What is hot work?
2. What three kinds of non-ionizing radiation do most welding processes produce?
3. How can noise harm a person?

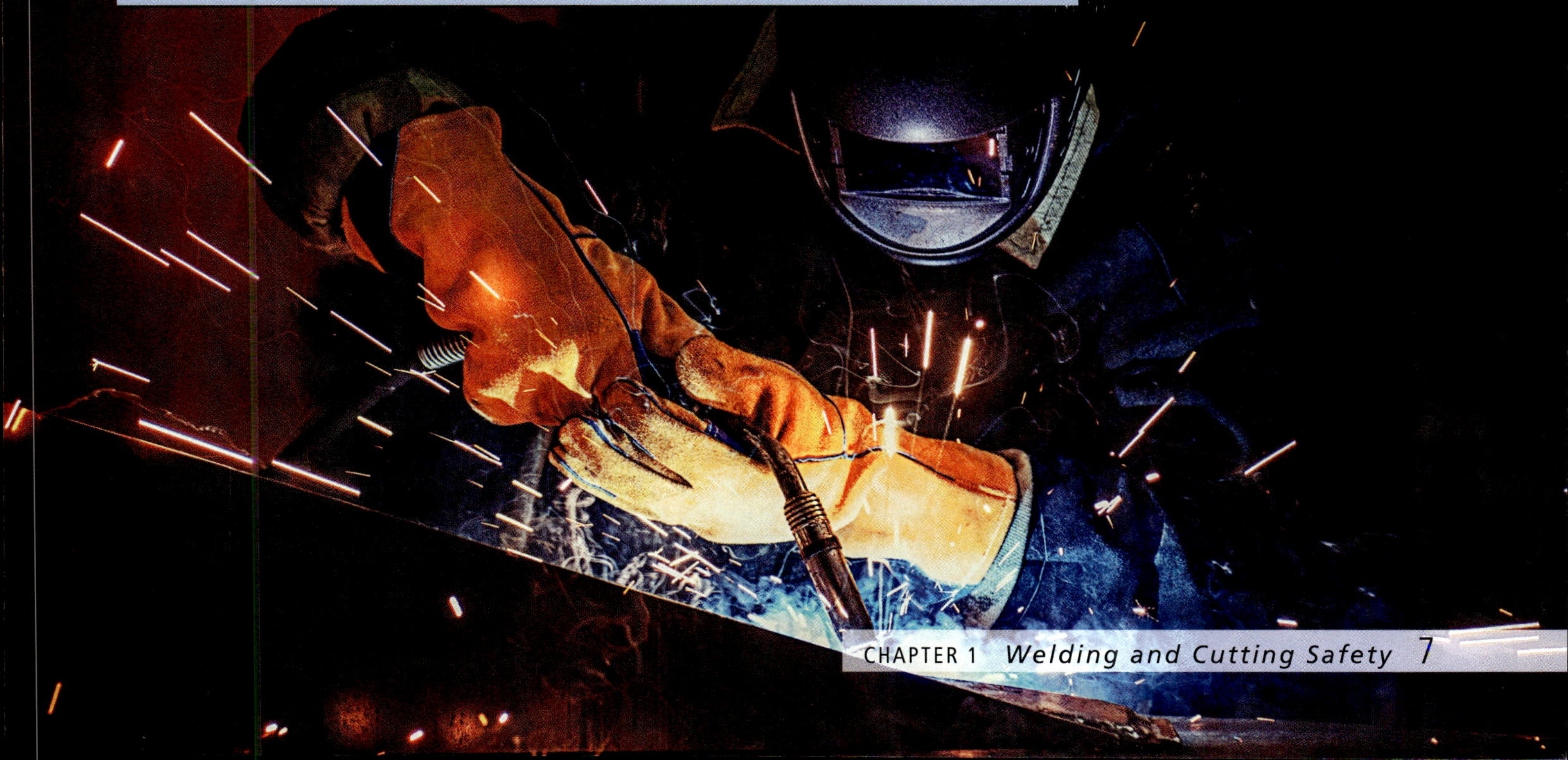

2 Work and Safety Clothing

In welding, clothing and footwear are an important part of **personal protective equipment** (**PPE**). There are two types of clothing: work clothing and safety clothing. Work clothes are those ordinarily worn to work and on the job; safety clothes are those that are added to work clothes to offer added protection from hazards on the job. Personal protective equipment also includes safety glasses, helmets, and other items specifically designed to help protect the welder and any observers from hazards associated with welding.

safety tip

Never wear polyester when welding or in an area where welding is taking place.

Work Clothing

Clothing should be made of wool, cotton, or a commercial fire-resistant material. Clothing made of polyester is completely unacceptable because it will melt when exposed to heat or sparks.

Clothes should be in good repair and should not have any frayed edges. These edges could catch fire if a spark comes into contact with the fraying.

safety tip

Ensure all combustibles are removed from clothing before welding, cutting, or grinding.

Pants Pants should not have cuffs, as they are potential collection places for sparks and splatter. Pants should be clean and free of frayed edges. Pants should be worn outside of boots to help prevent sparks and spatter from entering boot tops.

Shirts Although shirts should be buttoned to the top so that no skin is exposed to radiation, it is assumed that shirts will be worn under jackets when welding work is being done. If shirts have pockets, these should have flaps to prevent sparks that may pass to the shirt from the jacket.

safety tip

When buying work footwear, be certain it meets the minimum safety standards established by the American National Standards Institute (ANSI).

Footwear

The safety-conscious welder always wears boots or high-top shoes.

Boots Welders wear smooth leather pull-on boots with steel toes. If lace-up boots are worn, they should be fitted with bellow tongues to prevent sparks and slag from entering the boot and should have leather laces. If low-cut boots are worn, be sure to wear spats, discussed in Leather Accessories, to keep flying metal out of the boot.

Safety Clothing

Safety clothing gives workers required protection from hazards on the job. Safety clothing can be made of leather or cotton treated with flame retardant. Some safety clothing is available in both leather and treated cotton. Because of its durability and relative fire resistance, leather is often used for safety clothing. See Figure 3.

Jackets Cotton jackets are cooler in the summer, but when welding overhead leather jackets offer better protection. Jackets help protect the welder's upper body from slag, sparks, and other welding debris. Care should be taken to keep the jacket clean and free of oil and grease. Jackets should be buttoned to the top so that no skin is exposed to radiation.

Sleeves **Sleeves** protect a welder's arms from flying spatter and sparks. These "shirtless" sleeves can be worn over existing clothing and in combination with other protective gear. When welding overhead, leather sleeves offer very important protection against falling welding debris.

Gloves Leather welding gloves come in a variety of styles to suit different types of welding. Leather gloves should come with a gauntlet to help protect the welder's wrist area. Leather work gloves rather than leather welding gloves are used in work situations outside of welding.

Leather gloves should be kept dry. Damp or wet leather gloves do not protect the wearer from electrical shock. If the gloves are torn or otherwise sustain damage where skin can show through, the gloves should be replaced.

Spats Leather **spats** are shields that slip around a welder's ankle, lower shin, and foot area. Leather spats prevent spatter and sparks from burning the skin or clothing in these sensitive areas.

Aprons Aprons protect the area from the chest to the thighs from heat, spatter, and sparks. Aprons, like other leather garments worn while welding, should be kept clean of oil and grease. If a tear or other damage

safety tip

Position the hands so the glove gauntlet opening is facing down whenever possible. This will keep hot sparks from entering the glove.

Figure 3
Welder with protective gloves, jacket, beanie, apron, helmet, safety glasses, and boots

develops, the apron should be replaced. Aprons are effective protection when welding in any position. See Figure 4.

Bibs **Bibs** are part of a leather ensemble to protect the upper torso. This is especially critical when welding overhead. In this position, it is much more likely that sparks and splatter will land on the welder.

Chaps **Chaps** protect a welder's legs. Chaps are joined leggings worn over trousers that can be useful when welding is performed while standing. If welding is done while sitting or kneeling, an apron offers more protection from flying debris.

Welding Caps or Beanie Caps or beanies protect a welder's head from burns. One or the other should always be worn under a welding helmet for protection. A welding beanie is shown in Figure 5.

productivity tip

Select welding clothing that allows free range of motion. If clothing restricts motion, the welder's productivity is lower.

Figure 4
Welder wearing an apron

Figure 5
Welding beanie

Self Check

1. Why is polyester completely unacceptable for welding clothing?
2. Why are welders' accessories made of leather?
3. What kind of gloves should be worn when welding, and why?

3 Helmets: The Critical Safety Factor

Since ultraviolet and infrared rays from welding can harm the eyes even when standing 50′-0″ away and damage the skin within 20′-0″ away, it is very important to wear the proper safety gear when working on or near welding. The piece of PPE that protects the welder's head and eyes is the **welding helmet**. Sunglasses are never an acceptable welding shade. Anything not specifically designated as a welding shade is not appropriate for welding and should not be used.

Whatever the individual helmet preference, the helmet must be fully adjustable—away from the face and as to the angle it fits on the welder's head in relation to the eyes. This is very important, as safety glasses with side shields must always be worn under the welding helmet.

Figure 6
Welding helmet fitted to a hard hat

safety tip

Welders with long hair must tie it behind their necks.

Helmet Construction

Helmets are made out of heat-resistant materials, such as various types of plastic or even leather. Helmets feature an adjustable headband and crown, a foam or other absorbent material insert, ratchet controls to control the visor lift rate, and on some models, a lens-darkening adjustor and solar cell. Some helmets also feature a place to attach a hard hat. See Figure 6. These are called hard hat lugs.

Helmet Styles

There are several helmet styles available, each suitable for the type of welding being performed or the welder's personal preferences. All helmets have **visors**, the usually shaded window through which the welder can see. Whatever the preference, it is very important to wear a helmet while welding.

Fixed Shade The fixed shade is a one-piece helmet. It has an adjustable free-floating headband with an adjustable crown. See Figure 7. As the name indicates, the shaded lens cannot be moved independently of the hood. The entire helmet can be pivoted at the head gear to inspect work. These hinges can be adjusted tighter or looser, depending on the operator's preference.

Figure 7
Fixed shade helmet

safety tip

Safety glasses with side shields must be worn underneath a helmet at all times.

Figure 8
Auto-darkening helmet

Auto-Darkening Helmet An auto-darkening helmet allows the welder to see the work piece without flipping up the helmet. The helmet has adjustable shade settings depending on the brightness of the work. These helmet features vary by level of expense and function desired—number of sensors, type of batteries, degrees of shading, and size of the lens. See Figure 8.

Powered Air-Purifying Respirator Helmet (PAPR) This is a high-efficiency respiratory protection system for welders that has a rechargeable battery. The respiratory filters are on a belt around the welder's waist or in a backpack with a hose connecting to the welding helmet to provide fresh filtered air to the welder's face. The welding helmet is also an auto-darkening helmet; rather than the helmet having to be removed, it opens in the front to reveal a clear face shield. Figures 9 and 10 show PAPR equipment.

Types of Lenses

Helmets come with a variety of lenses. These lenses vary in size, material, and degree of shading. They are made of polycarbonate plastic or glass.

There are several parts to every lens. On most helmets, there is a shaded filter lens, the protective outer lens, and a secondary protective inside lens. The outer and inner lenses are made of optical plastic or glass and help protect the more expensive filter from welding sparks, slag, and heat.

Polycarbonate Plastic Polycarbonate plastic lenses come clear and shaded. These lenses protect both sides of the more expensive glass filter lenses from damage. A variety of shade levels are available. The welder should use the recommended shade based on the work being performed. See Table 1 in Degrees of Shading. When these lenses become scratched and damaged from extensive use, they should be replaced.

Figure 9
PAPR equipment

Glass Glass lenses provide good protection and come in both clear and shaded styles. Most helmets come with two or three layers of glass. The outer glass lens protects the shaded lens from damage. The welder should use the recommended shade for the inner lens, based on the work being performed. See Table 1, Recommended shading rating for various welding processes, on the following page. When these lenses become scratched and damaged from extensive use, they should be replaced.

Gold Gold lenses are available in glass or polycarbonate plastic. The gold shade allows the welder to see the arc in natural color while reflecting 90 percent of the welding heat.

Magnifiers Also known as cheaters, these lenses can be inserted into the regular lens to magnify work. Magnifiers are used to correct for poor eyesight. Like shade ratings, the higher the number, the higher the magnification.

Automatic Lenses One of the latest developments in lens technology is the auto-darkening lens. These lenses darken within $\frac{1}{25,000}$ of a second when an arc is detected. The lens has either a predetermined shade limit or can be manually adjusted to suit the type of welding being performed. These lenses are battery-powered and often have a solar battery booster.

Figure 10
PAPR system

Lens Installation Sequence

Since the welding helmet lenses need to be periodically replaced, it is important to know how to replace them without damaging the helmet or decreasing its effectiveness. The correct sequence in which the lenses should be inserted, working from the outside in, is listed below.

- clear outside lens
- outermost lens gasket
- shaded lens
- innermost lens gasket
- clear inside lens
- retainer

productivity tip

Automatic lenses allow the welder to quickly position the welding arc and reduce the need to raise and lower the welding hood or lens. This leaves more time for welding.

Self Check

1. Why is it important to wear a helmet?
2. In what circumstance might a magnifier lens be used?
3. Can the eye see the arc striking when using an auto-darkening lens?

Degree of Shading

It is important to use a welding lens that is appropriately shaded for the work. A lens that is too dark can impede work and cause eye strain, while using a lens that is too light can result in eye damage.

Lens shading is measured by **shade rating**, with number ratings ranging from 2 to 14, with 2 being the weakest and 14 the strongest. Shaded lenses provide excellent protection for vision. A welder can choose a darker shade lens for work being done but not lighter.

The Occupational Safety and Health Administration (OSHA) publishes a recommended shade rating table. Please refer to this information when choosing shade strength. Table 1 shows recommended shade rating for various welding processes. Both the Occupational Safety and Health Administration (OSHA) and the American Welding Society (AWS) publish shade rating tables.

TABLE 1 ***Recommended shade rating for various welding processes***

Shade	Recommended uses
2	Reflected glare and low temperature furnace work
3	Light brazing and lead burning
4	Acetylene burning and brazing
5	Light acetylene welding and cutting
6	Standard shade for acetylene gas welding
8	Heavy acetylene welding, electric arc cutting, and welding up to 75A
10	Electric arc cutting and welding between 75A and 250A
12	Electric arc cutting and welding above 250A
14	Carbon arc cutting and welding

Summary

Welding safety involves protecting people and the environment from the by-products of welding, such as heat, sparks, radiation, fumes, smoke, and noise. Using personal protective equipment is an important part of staying safe while welding. It is also important to be aware of the environment: know what is flammable in the work area, understand the ventilation situation, and be aware of others working nearby. Clothing should be carefully chosen and appropriately worn to provide the most protection possible. Careful consideration should be paid to protecting eyes from radiation and ears from noise.

Chapter Review

Show your understanding of the information in this chapter by answering the questions and filling in the blanks below.

1. Hot sparks can fly up to ____________________ feet away.

2. Which of the following is not considered radiation?
   a. ultraviolet light
   b. infrared light
   c. visible light
   d. gold lens

3. The combination of ____________________ and ____________________ provide the best hearing protection.

4. The least safe material to wear while welding is ____________________.

5. What are the safest shoes to weld in?
   a. tennis shoes
   b. sandals
   c. leather pull-on boots with steel toes
   d. hiking boots

6. ____________________ and ____________________ ____________________ are two properties that make leather a good protective clothing choice for welders.

7. What do spats help protect?
   a. head
   b. arms
   c. fingers
   d. feet

8. It is acceptable to have exposed pockets on a leather welding jacket. (True; False)

9. What piece of personal protective equipment helps protect wrists?
   a. gloves
   b. spats
   c. chaps
   d. aprons

Chapter Review

10. The bib is part of a leather ensemble to protect the upper torso. (True; False)

11. Welding helmets are made of ____________________ or ____________________.

12. Lenses are made of ____________________ or ____________________ ____________________.

13. The higher the shade number, the lighter the shade. (True; False)

14. When a lens is damaged, it should be replaced. (True; False)

15. What is the benefit of using a gold-colored lens?
    a. small things look larger
    b. large things look smaller
    c. color within the welding field remains true
    d. they protect from arc flash

CHAPTER 2

Oxyfuel Safety

Oxyfuel cutting requires cylinders to store and dispense fuel gas and oxygen. Fuel gas can be acetylene, propane, natural gas, or MAPP® gas. The cylinder and hose requirements for each of these gasses are different. Knowing how to safely handle and work with cylinders is essential for effective welding.

Contents

Key Terms

check valve device that restricts movement of gas to one direction only

cylinder used to contain fuel gas and oxygen under pressure; usually made of steel

flashback arrestor safety device that prevents flame from traveling upstream of the cutting tip

fuel gas gas used together with oxygen to preheat the metal to be cut; usually propane, acetylene, MPS (MAPP®), or natural gas

manifold chamber used in conjunction with a single-stage regulator to reduce initial gas pressure before it reaches the regulator

regulator device that converts gasses under high pressure to lower pressures suitable for cutting; comes in single-stage or two-stage varieties

Objectives

- Describe aspects of cylinder safety, including transportation, storage, and size.
- List types of hoses and describe how they connect to cylinders.
- Demonstrate an understanding of the proper amount to open valves on cylinders.

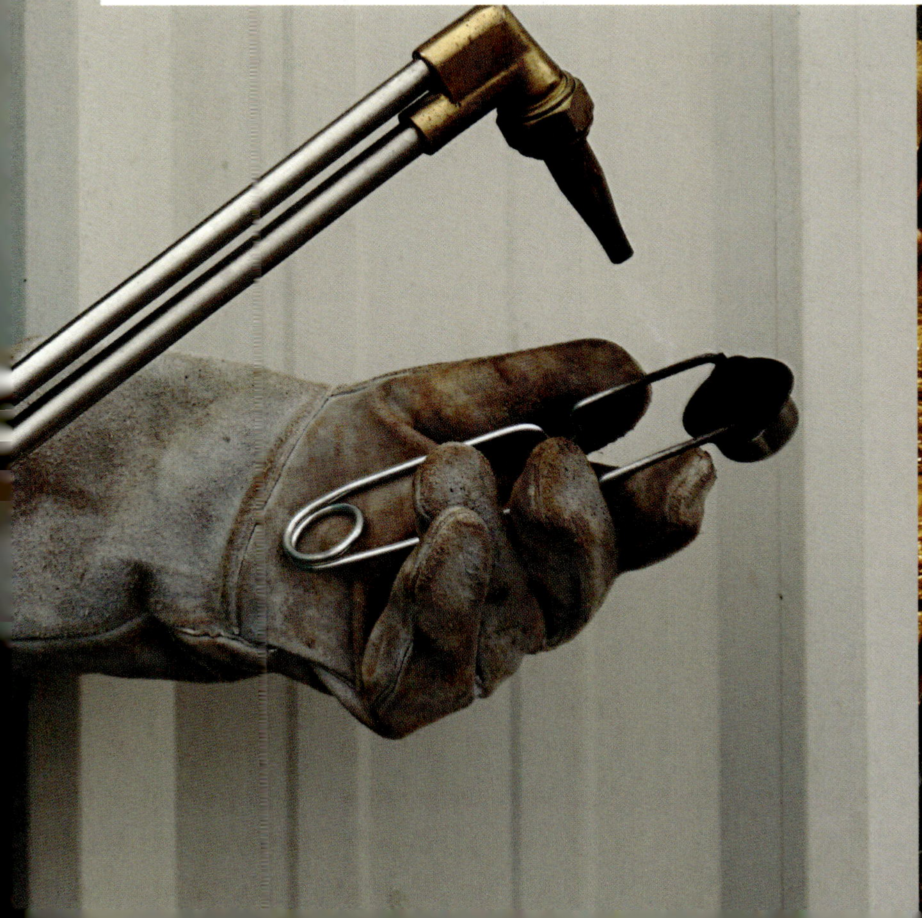

1 Cylinder Safety

There are numerous federal and state regulations governing all aspects of working with gas cylinders. **Cylinders** contain fuel gas under pressure. A **fuel gas** is a gas used together with oxygen to preheat the metal to be cut. The information below represents general gas cylinder safety and handling practices. For specific rules and regulations, consult the local regulatory agency.

Securing and Transporting Cylinders

When moving cylinders from job to job, special care must be taken to protect the cylinders, the people transporting them, and the general public.

Before moving a cylinder, remove the regulator and put the valve protection cap on. This should be done unless a transport that is specifically designed to transport cylinders is being used. Always secure cylinders in an upright position where they cannot fall over or be knocked over accidentally. The following steps are recommended safety procedures:

- Do not use the valve protection cap to lift a cylinder.
- If the valve under the protection cap becomes frozen, use warm, not boiling, water to thaw the valve. Never use a bar to force the cylinder open.
- A suitable cylinder cart, chain, or other steadying device must be used to keep cylinders from being knocked over while in use.
- When hoisting cylinders, secure them on a cradle, sling board, or pallet. Do not use magnets or choker slings.
- Move cylinders by tilting and rolling them on their bottom edges. Do not drop the cylinders or let them strike anything.
- When transporting cylinders in a vehicle, they should be secured in an upright position.
- When work is finished, when cylinders are empty, or when cylinders are moved at any time, close the cylinder valve, even when cylinders are on a portable cart.

Storing Cylinders

It is important to store cylinders correctly. Incorrectly stored cylinders can be damaged, their contents lost, or a fire hazard created. The following are recommended ways to correctly store cylinders.

- Always secure cylinders in an upright position where they cannot fall over or be knocked over accidentally.

- Keep cylinders far enough away from the welding or cutting operation so that sparks, hot slag, or flame cannot reach them. If this is impractical, use a fire-resistant shield such as a fire blanket.
- Keep cylinders away from any source of artificial heat. Remember the 35′-0″ rule: sparks and molten metal can travel up to 35′-0″ horizontally from the work area!
- Do not place cylinders where they can become part of an electrical circuit; they must be a minimum of 10′-0″ away from an open power source.
- Do not accidentally touch the cylinder with an electrode.
- Always keep cylinders with the valve end up when in use or in storage. Never lay cylinders flat.
- Do not place cylinders in a confined space.
- Do not store oxygen and fuel cylinders together. They must be separated by at least 20′-0″, or separated by a 1.5-hour-rated, noncombustible, 5′-0″ tall barrier.

...choose the right fuel cylinder for the job...

Choosing the Right Cylinder Size

Fuel cylinders come in various sizes. It is important to choose the right fuel cylinder for the job so that no time will be wasted changing an empty bottle for a full one. It takes a great deal of time to change the bottle during a work procedure. The size should be based on the amount of time it will take to perform a job. As a rule, the fuel cylinder should be of a size so that no more than 1⁄10 of the capacity of the cylinder will be used per hour during intermittent use or 1⁄15 of the capacity of the cylinder per hour during continuous use.

Oxygen cylinders also come in various sizes. The oxygen cylinder should be large enough to finish the cutting task at hand.

Figure 1 shows the various capacities in which cylinders are available. The larger cylinders are constructed of steel. The smaller cylinders are

productivity tip

Choose the right size cylinder before starting to work so that an empty cylinder does not have to be swapped for a new, full one. It can take a lot of time to perform the swap.

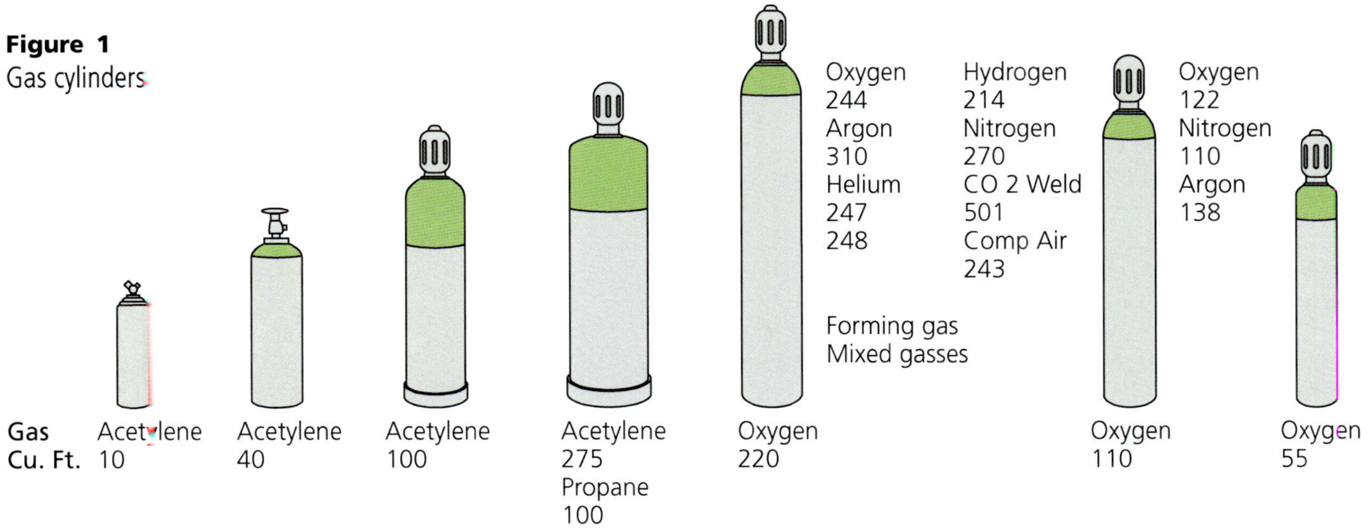

Figure 1
Gas cylinders

Figure 2
Check valve for an oxyfuel system

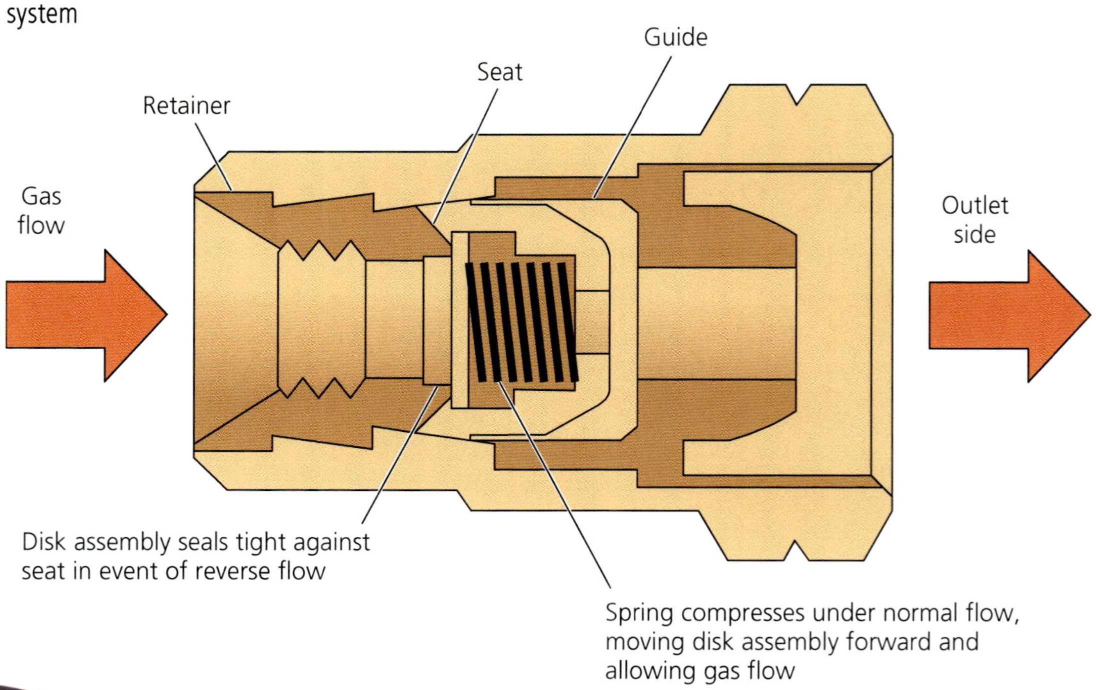

made of aluminum. These smaller, lighter cylinders are typically used for medical gasses only. They are not usually appropriate for welding gasses.

Additional Cylinder Safety Practices

Cylinders should always be kept upright, when storing, moving or in use. Acetylene cylinders have a honeycomb porous material with acetone inside. Acetylene gas is a compound of carbon and hydrogen and is *volatile* at more than 15 pounds per square inch (psi). If the welder attempts to use an acetylene cylinder after the cylinder has been on its side, the acetone, which is *corrosive*, will adversely affect the cutting torch parts and the welder.

If a cutting torch tip is blocked, if there is an excess or lack of gas pressure, or if an unsafe startup or shutdown procedure has been followed, gas can flow backward through the system. There are two devices welders use to prevent this from happening: check valves and flashback arrestors.

Check Valves A **check valve** allows gas to flow in one direction only. These valves prevent the reverse flow of gasses. If check valves were not used, it would be possible for fuel gasses to enter the oxygen tank or oxygen to enter the fuel tank, a condition called backflow. Figure 2 shows how a check valve works.

Flashback Arrestors A **flashback arrestor** is another important safety device. During a flashback, flame and gas from the torch pass upstream into either gas system: fuel or oxygen. Flashback arrestors prevent the gas from flowing backward and the flame from traveling upstream of the arrestor. The use of a flashback arrestor is encouraged to prevent this dangerous situation from occurring. Figure 3 illustrates the parts of a flashback arrestor.

safety tip

Local regulations frequently require the use of flashback arrestors. This eliminates backflow of gasses into bottles that would cause an explosion.

Self Check

1. When should valve protection caps be used?
2. When is it acceptable to store cylinders on their sides?
3. How is fuel cylinder size determined?

Figure 3
Flashback arrestor for an oxyfuel system

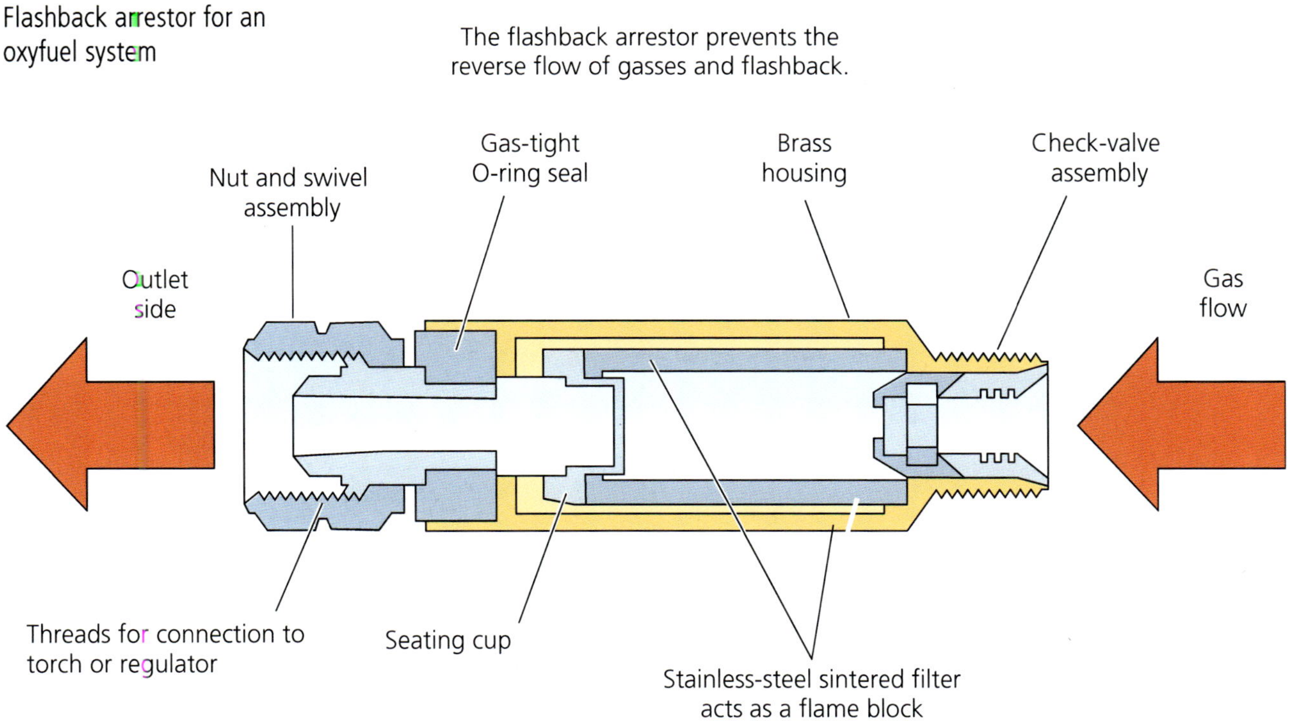

2 Hose Safety

It is important to understand all parts of the oxyfuel rig thoroughly so that mistakes are not made in setting up the equipment. One such part involves the hoses that connect the cylinders to the torch.

Hoses are available in three sizes—3/16″, 1/4″, and 3/8″ inner diameter (i.d.). Grade T hose is for fuel gas, including acetylene. This replaces former grade M. Grades R and RM are for use with acetylene only. Hoses are usually colored to show which gas is being carried. The oxygen line is green; the fuel line is usually red.

The cylinder regulators and torch head are usually connected by double-line rubber hoses, known as twin weld.

trade tip

Do not to mix hoses when oxyfuel welding. Always use the same hose for oxygen and always use the same hose for fuel. If hoses are switched, gas impurities could affect the welds.

safety tip

Small sections of the hoses may be taped together to avoid tangles. However, do not cover the hoses with tape because this prevents them from being inspected.

Hose Connections

It is important to use only the correct hose connectors on the regulator and the cutting torch. To keep from mixing up the hoses, connectors, and regulators, there are different systems for oxygen and acetylene fittings.

The oxygen hose connectors consist of a barbed nipple that is pressed into the hose and clamped with a *swaged collar* and screws into the cutting torch or regulator using a nut. The oxygen nut has a right-hand thread pattern.

The fuel connectors use a similar setup; however, instead of a right-hand thread pattern, fuel connectors have a left-hand thread pattern.

use only the correct hose connectors...

3 Using Cylinder Regulators Safely

The oxygen and fuels used during oxyfuel welding are under tremendous pressure inside the cylinders. In the case of oxygen, its pressure is about 2,200 psi in its cylinder. If the oxygen were just released at that pressure, no welding could be performed. Therefore, regulators are used to reduce the pressure to a usable amount.

A **regulator** reduces the pressure that the fuel and oxygen is released at and smoothes the delivery to help keep the welding flame even. Each gas cylinder comes with its own regulator. Figure 4 illustrates two regulators.

Types of Regulators

There are two basic types of regulators: single-stage regulators and two-stage regulators. Each regulator works through gas pressure, screws, springs, and diaphragms.

Single-Stage Regulators Single-stage regulators are most often used with a separate manifold. In this system, the **manifold** reduces the initial gas pressure before it enters the regulator system. Through the manifold chamber, the initial gas pressure is reduced to a fixed amount determined by screws located on the manifold.

Two-Stage Regulators The two-stage regulator works without a manifold. Instead, the regulator has two chambers: a high-pressure chamber and a reducing chamber. The pressure in each of these chambers is set by adjusting screws.

trade tip

The nuts for torch fittings provide a clue as to which way it threads. If the nut has a groove around it, it is a left-hand thread and is therefore a fitting for a fuel hose. If the nut is smooth, it indicates a right-hand thread and is the fitting for the oxygen hose.

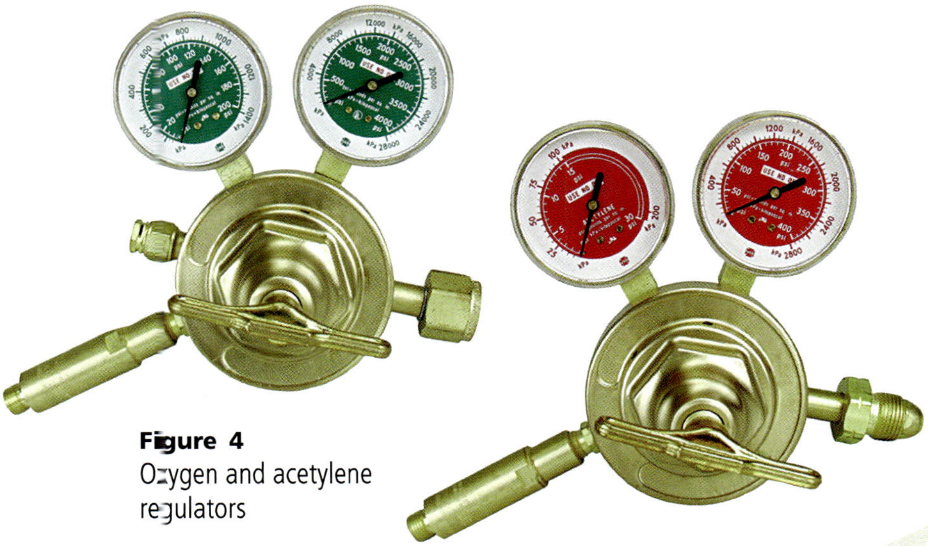

Figure 4
Oxygen and acetylene regulators

Figure 5 illustrates the inner workings of the single-stage and the two-stage regulator.

Self Check

1. Is it easy to mix up the oxygen and fuel hose connectors?
2. When is it acceptable to use a single-stage regulator?

Attaching Oxygen Cylinder Regulators Safely

Before connecting a regulator to an oxygen cylinder valve, the valve should be *cracked*, that is, it should be opened slightly and closed immediately. This action will clear the valve of dust or dirt that might otherwise enter the regulator. The person cracking the valve should not stand in front of the cylinder valve.

Other safety measures include the following.

- Visually inspect regulators and gauges for visible damage to the gauge and fitting seats.
- When attaching oxygen cylinders to the regulator, never overtighten the connections as this can damage brass fittings on the regulator.
- Never oil the oxygen regulator or the cylinder valve. In the presence of high-pressure oxygen, oil and grease can spontaneously combust.
- Always check that the regulator's pressure adjustment screw is backed out to its fullest extent while remaining attached to the regulator prior to opening the cylinder valve.
- When opening the cylinder, the welder should stand on the opposite side of the cylinder from where the regulator is pointing. If the regulator has a damaged diaphragm or damaged threads on the adjustment screw, the 2,250 pounds of pressure can cause the regulator to fly into pieces, turning those pieces into dangerous projectiles.
- Finally, when properly positioned on the side away from the direction the regulator is pointing to, the oxygen cylinder valve should be opened slowly and completely to avoid damaging the regulator.

trade tip

Because oxygen cylinders are typically at 2,200 psi, they have a double-seat valve assembly. Due to the double-seat valve, oxygen cylinder valves and all other high-pressure gas cylinder valves must be fully opened or fully closed.

Attaching Fuel Cylinder Regulators Safely

Before connecting a regulator to a fuel cylinder valve, the valve should be cracked, that is, opened slightly and closed immediately. This will clear the valve of dust or dirt that might otherwise enter the regulator. The person cracking the valve should not stand in front of it. The valve of a fuel gas cylinder should not be cracked where the gas can reach welding work, sparks, flame, or other possible sources of ignition.

Never oil the fuel regulator or the cylinder valve. In the presence of certain gasses, oil and grease can spontaneously combust.

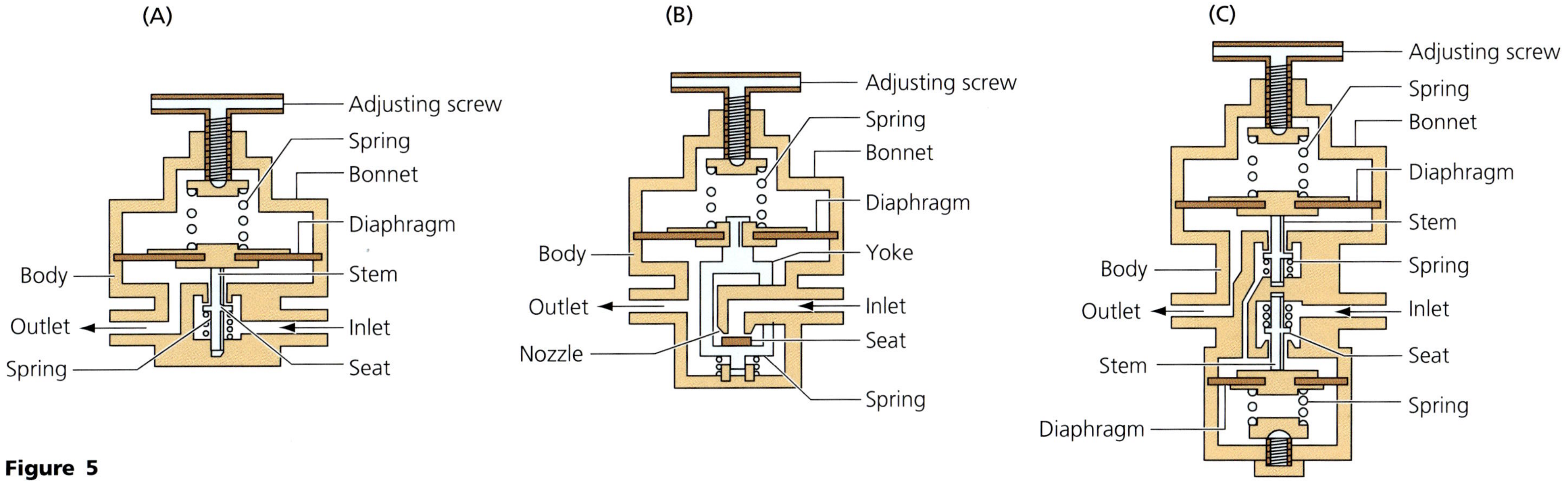

Figure 5
Typical gas regulators
(A) Single-stage stem type
(B) Single-stage nozzle type
(C) Two-stage

Once the fuel cylinder valve has been cracked and the regulator has been attached, the cylinder valve should be opened slowly to prevent damage to the regulator. For safety purposes, never open fuel cylinder valves more than 1 ½ turns.

If a special wrench is required to open the fuel cylinder valve, that wrench should be left in position on the valve stem in case the gas flow needs to be quickly shut off. Never store anything on top of the cylinder. Placing objects there may interfere with the regulator operation or create a safety hazard if the gas needs to be quickly shut off.

safety tip

Never open fuel cylinder valves more than $1\frac{1}{2}$ turns.

safety tip

Never store anything on top of the cylinder.

safety tip

Back off regulators after every shift to stop valves from getting sticky.

Checking for Leaks

Before the torch can be turned on, the connections must be checked for leaks. If soapy water is not available on the jobsite to check for forming bubbles, there is a standard procedure frequently used.

PROCEDURE

How to Check for Leaks

1. Once the regulator has been properly attached to the cylinder and before touching the regulator adjusting screw, crack the oxygen bottle and watch the high-pressure gauge. The gauge should now reflect the amount of pressure in the cylinder. If the pressure reading starts to drop, there is a leak in the connection between the gauge and the cylinder. That fitting must be tightened, and the high-pressure gauge checked again to make sure the pressure is holding.
2. Once the torch head and hose are connected, open the oxygen cylinder and adjust the regulator adjustment screw until the desired pressure is shown on the low-pressure gauge.
3. Note the pressure on both gauges, shut the cylinder valve, and watch the low-pressure gauge to see if the pressure drops. If the pressure has gone down, a connection between the manifold and the torch head—or one of the valves in the torch head—is leaking. In this case, tighten the connections and repeat the check. If both gauges hold steady pressure, there are no leaks.

trade tip

Fuel cylinder valves have a single-seat valve assembly because their contents are not under a lot of pressure.

trade tip

Only open acetylene cylinders from $\frac{3}{4}$ to $1\frac{1}{2}$ turns.

Summary

Following good safety procedures when cutting is very important. The gasses used in oxyfuel cutting should be treated with respect. Always check hoses, nozzles, and cylinders before starting any cutting job. Always secure cylinders when transporting them from job to job. Equipment that is in good working order and that is operated safely increases job efficiency.

Chapter Review

Show your understanding of the information in this chapter by answering the questions and filling in the blanks below.

1. A welder is working on a big job 25 miles from the shop. What is the first thing he or she should do to begin safely moving the oxyacetylene rig?

   ____

   ____

   ____

   ____

2. Cylinders may never be stored on their sides. (True; False)

3. The ____________________-____________________ regulator works without a manifold.

4. ____________________-____________________ is the thread orientation for oxygen cylinders.

5. The oxygen hose is normally green. (True; False)

6. A single-stage regulator should be used only when there is a(n) ____________________.

7. How much should you open a fuel cylinder valve?
   a. ¼ to ½ turn
   b. ⅓ to ⅔ turn
   c. not more than 1 ½ turns
   d. ½ to ¾ turn

8. What could possibly happen if you mixed up the hoses or connectors of the oxygen cylinder and the fuel cylinder?

   ____

   ____

   ____

   ____

9. Why is it important to "crack" the oxygen cylinder before use?

____

____

____

____

10. A check valve prevents the ____________________ flow of gasses.

CHAPTER 3

Oxyfuel Cutting

A widely used method for cutting metal is oxyfuel cutting (OFC). Several different fuel gasses can be used with oxygen. The type of fuel gas to be used is determined by the type of metal being cut. There are several types of torches that can be used. The type of torch chosen should be best suited for the job at hand.

Key Terms

acetylene popular fuel gas for cutting

cutting torch mixes a fuel gas for heating flames and a stream of pure oxygen to cut metal

fuel gas used with oxygen to create cutting flame; heats the metal enough to produce the desired chemical reaction with oxygen

ferrous metal metal that contains iron

kerf width of the cut

MAPP® brand of acetylene-based fuel gas used in oxyfuel cutting

MPS cutting gas acetylene-based fuel gas used in oxyfuel cutting

natural gas fuel gas used in oxyfuel cutting

oxyfuel cutting (OFC) using oxygen and a fuel gas to cut a ferrous metal

propane fuel gas used in oxyfuel cutting

Contents

Objectives

- Demonstrate an understanding of and describe how oxyfuel cutting works.
- Identify types of fuels used for cutting and list their advantages and disadvantages.
- List the different torches used in cutting and their uses.

1 Personal Protection Equipment

It is important to wear approved clothing and accessories when cutting with an oxy-fuel torch.

- Wear dry gloves that have no tears or holes.
- Wear protective garments such as leather gloves, a heavy shirt, cuffless pants, high shoes or pull-on boots or boots with leather laces, and a cap.
- Wear approved safety goggles with side shields in addition to wearing a shaded face shield. Either the goggles or the face shield must have a minimum shade value of 5.
- Use approved helmets that protect the face, neck, and ears.
- Wear leggings, aprons, sleeves, shoulder covers, and bibs.

Wear approved safety goggles with side shields...

2 What Is Oxyfuel Cutting?

The process called **oxyfuel cutting** (**OFC**), uses oxygen and some other fuel gas to cut **ferrous metals**. Ferrous materials are those metals that contain iron. It is the chemical reaction that takes place between ferrous metal and the oxygen present in oxyfuel welding that enables the metal to be cut. If there is no iron in the metal, that is, if it is a *nonferrous* metal, the chemical reaction described below will not take place.

Ferrous metals have an affinity for oxygen. Think of an old rusted engine that has been sitting out behind the shed, exposed to the elements. The iron present in that engine has reacted with the oxygen in the air to form an iron oxide—rust. This rusting process can take a long time, as the oxygen content in the air is only about 20 percent.

This process works to our benefit in oxyfuel cutting. When oxyfuel cutting is introduced to a metal that contains iron, the heat produced from the flame raises the metal to 1,500° F to 1,600° F, a cherry red color. As part of the cutting process, oxygen is blasted onto the hot metal. This causes the metal to burn; in essence, it burns just like kindling wood burns once it reaches the right temperature.

When wood burns, it gives off smoke as one by-product of the chemical breakdown induced by heat. Metal also has a by-product when it burns. While there is some smoke, the main by-product of burning metal is iron oxide—rust. However, the melting point of iron oxide, or rust, is below that of the steel. The heat from the burning steel and the oxyfuel torch is enough to melt the rust as soon as it forms. The melted iron oxide forms a *slag* that runs off of the metal being worked, exposing more metal. Therefore, clean cuts can be made with oxyfuel flame.

Looked at in this way, oxyfuel cutting is simply very rapid rusting. It is therefore considered to be a chemical process, unlike arc welding which is considered a mechanical process. Oxyfuel cutting utilizes a **fuel gas** in addition to oxygen. The type of fuel gas depends on the material being cut. Remember though, it is the oxygen portion of the rig that is actually doing the cutting. The fuel gas merely heats up the metal enough so that it reacts with the oxygen.

Self Check

1. What chemical process happens to ferrous metal when it is oxyfuel cut?
2. What are some by-products of oxyfuel cutting?

3 Cutting Fuels

As mentioned, cutting fuels are used in conjunction with oxygen to create a heat source strong enough to melt metal containing iron oxide.

There are several types of fuel gasses available to choose from, and each has its strengths and weaknesses. The properties of the most common gasses are listed in Table 1.

Each of these gasses has advantages and disadvantages. Each also uses different torch tips. Understanding how each gas performs will enable the welder to make an educated choice on which gas to use for a particular job.

Acetylene Gas

By far the most popular fuel gas today, **acetylene** is used for both welding and cutting purposes. Most other gasses are compared against acetylene with regard to performance and cost.

Advantages An oxyacetylene flame has reached approximately 5,600° F, making it the hottest fuel available. However, acetylene's average *British thermal unit*, (Btu), a measure of heat production, is only 1,450 Btu. While

TABLE 1 ***Cutting fuel gas chart***

	Acetylene	Methylactylene-propadiene* MAPP®	Propane*	Propylene	Natural gas (methane)
Neutral flame temperature, °F	5,589°	5,301°	5,250°	4,579°	4,820°
Oxygen fuel combination ratio	2.5:1	4.0:1	4.5:1	5.0:1	2.0:1
British thermal units (Btus) per cu. ft.	1,470 Btus	2,460 Btus	2,400 Btus	2,243 Btus	1,000 Btus
Burning velocity in oxygen	22.7 ft./sec.	15.4 ft./sec.	15.0 ft./sec.	12.2 ft./sec.	15.2 ft./sec.
Flammability limits in oxygen**	3–93%	2.5–60%	2.5–60%	2.4–57%	5–59%
Flammability limits in air	2.5–80%	3.4–10.8%	2.0–11.1%	2.3–9.5%	5.3–14%
Specific gravity (air = 1@60° F)	0.91	1.48	1.48	1.52	0.62
Maximum allowable regulator pressure, psi	15	150	150	150	Line

* MAPP® and propane fuels are heavier than air and will collect in low areas such as basements. The accumulated fuels are very toxic and highly flammable. Always use extreme caution when using these fuels around holes or closed-in areas.

** Use caution when mixing fuel gasses with an oxidizer. When fuels are mixed in the proper proportion, they can become a potentially explosive material.

less than MAPP® or propane, the concentration of the Btus is greater than that of the other two gasses. Therefore it is more efficient—it heats up much faster than other fuels. Acetylene is ideal for cutting steel plate or for surfaces coated with grease, paint, or rust.

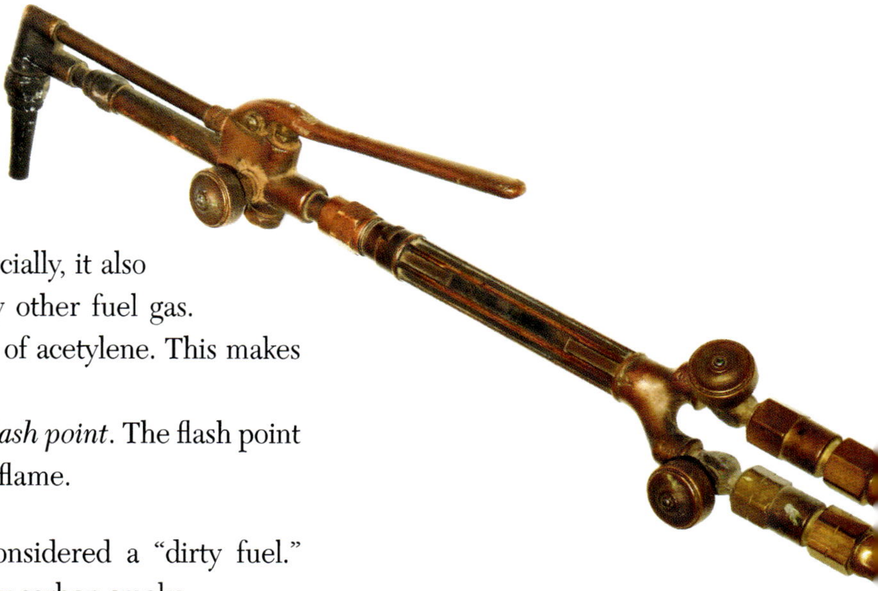

Not only is acetylene widely available commercially, it also uses less oxygen per cubic foot of fuel than any other fuel gas. Acetylene requires 1 unit of oxygen per 1.5 units of acetylene. This makes it an efficient gas.

Acetylene is easy to light, because it has a low *flash point*. The flash point is the amount of heat required for it to burst into flame.

Disadvantages Unfortunately, acetylene is considered a "dirty fuel." When acetylene burns, it burns with a heavy, sooty carbon smoke.

Acetylene is also quite volatile and can only be used up to a maximum 15 psi. Although it is packed with porous materials and filled half full with acetone that makes it safe, it still must be handled with great care due to its volatility.

Acetylene is also an expensive fuel gas compared to the other gasses.

Tip Styles The style and size of the tip used depends on the thickness of the metal being cut and the torch manufacturer. Oxyacetylene cutting typically uses a one-piece tip. Tips for acetylene are flat on the flame end.

Oxyacetylene tips range in size to produce a variety of flame sizes. The size of the tip is marked by numbers, commonly ranging from 000 to 15. The larger the number, the larger the flame hole and the greater the heat produced. The larger the tip size, the greater amount of gasses required.

See Figure 1 for types of special cutting tips.

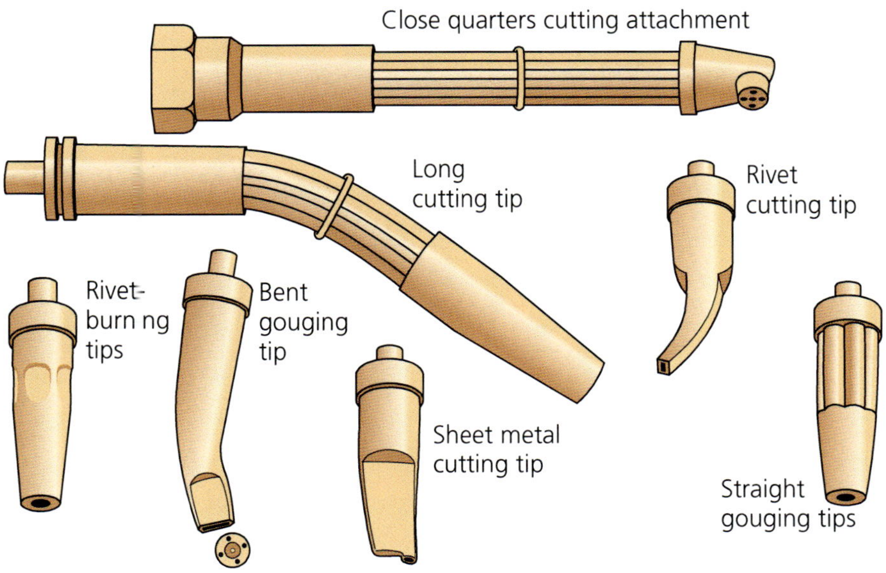

Figure 1
Special cutting tips

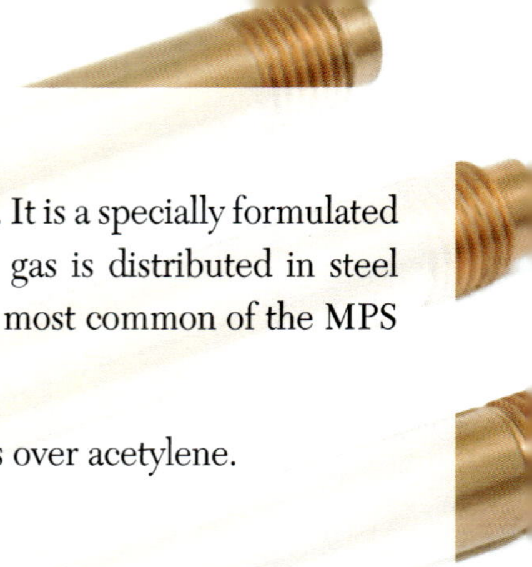

MPS Gas

MAPP® is the trade name of an **MPS cutting gas.** It is a specially formulated liquefied methylacetylene-propadiene gas. The gas is distributed in steel cylinders similar to acetylene cylinders. It is the most common of the MPS cutting gasses used in the United States.

Advantages MAPP® has two main advantages over acetylene.

- It is less expensive than acetylene.
- It is less volatile than acetylene.

MAPP® gas can be used at pressures of up to 375 psi. Therefore, a 120-pound cylinder contains as much gas as five 240-pound cylinders of acetylene. This reduces handling expense.

Because MAPP® has a strong odor, leaks are easily detectable—even at concentrations as low as 0.01%. It is about as volatile as propane or natural gas, both of which are much less volatile than acetylene gas.

Disadvantages While MAPP® can be used at a much higher psi than acetylene, its flame temperature is 5,300° F, about 300° less than acetylene's. Although MAPP®'s Btu is higher than acetylene (2,460 compared to 1,470), it takes longer to reach working temperature. Also, more fuel and oxygen is consumed while cutting with oxy-MAPP®. MAPP® gas requires 4.0 units of oxygen per 1 unit of MAPP®.

Tip Styles The typical tip for cutting with MAPP® has a flat surface on the flame end, just like the tip for acetylene. The tip can be a single tip or a two-piece tip. See Figure 1 for types of special cutting tips.

Propane Gas

The gas called **propane** is a *hydrocarbon gas* present in petroleum and natural gas. In liquefied state, it is transported and stored in steel cylinders ranging in size from 20 to 100 pounds. Propane is widely used for oxyfuel heating, cutting, *soldering*, and *brazing*. It has a distinct smell added to it during processing that makes detecting leaks easy.

Advantages One advantage of propane over acetylene is that it costs less. It is also less volatile than acetylene.

Disadvantages While propane has a Btu of 2,400, the flame is less concentrated than acetylene and therefore it takes longer to heat metal to working temperature. Propane gas requires 4.7 units of oxygen per 1 unit of propane.

Tip Styles Propane cutting tips are typically one- or two-piece units. These tips have a deeper recess or a cupped end to direct the flame. See Figure 1 for types of special cutting tips.

Natural Gas

A mixture of combustible hydrocarbons, especially *methane*, **natural gas** has many uses for oxyfuel cutting. Because it can be safely compressed to high pressures, it is a good gas for underwater cutting.

Advantages Natural gas is widely available and can be stored in large compressed tanks or hooked directly up to city gas services. It works well in production shops.

Natural gas is also less expensive than any of the other fuel gasses. Natural gas is safer to handle and less volatile. Natural gas requires two units of oxygen per one unit of natural gas, making it somewhat more efficient to use than MAPP® and propane.

Disadvantages At 918 Btu, natural gas does not pack the heating power of the other fuel gasses. The combined intensity of the flame is only 5,600 Btu, less than half of acetylene's. In addition, since natural gas is not available in portable cylinders, it cannot be moved from job to job.

Tip Styles Natural gas cutting tips are typically one- or two-piece units. These tips have a deeper recess or a cupped end to direct the flame. See Figure 1 for types of special cutting tips.

Self Check

1. What range of cutting tip sizes is available to welders?
2. What characteristics of acetylene make it a popular fuel gas for cutting?
3. Which cutting fuel is also widely used for household appliances?

4 Cutting Torches

The oxyfuel **cutting torch** is designed to supply a fuel gas for heating flames and a stream of pure oxygen to cut the metal. Figure 2 illustrates the inner workings of a typical cutting torch.

There are a number of types of torches available and each is suitable for a particular gas or job. See Figure 3.

Cutting torches can produce a variety of flames. See Figure 4 for a full range of flame types.

Fuel Lines

It is important to use only the correct hose connectors on the regulator and the torch.

Hose Connectors To keep from mixing up the hoses, connectors, and regulators, there are different systems for oxygen and fuel gas fittings.

- The oxygen hose connectors consist of a barbed nipple that screws into the torch or regulator using a nut. The oxygen nut has a right-hand thread pattern.

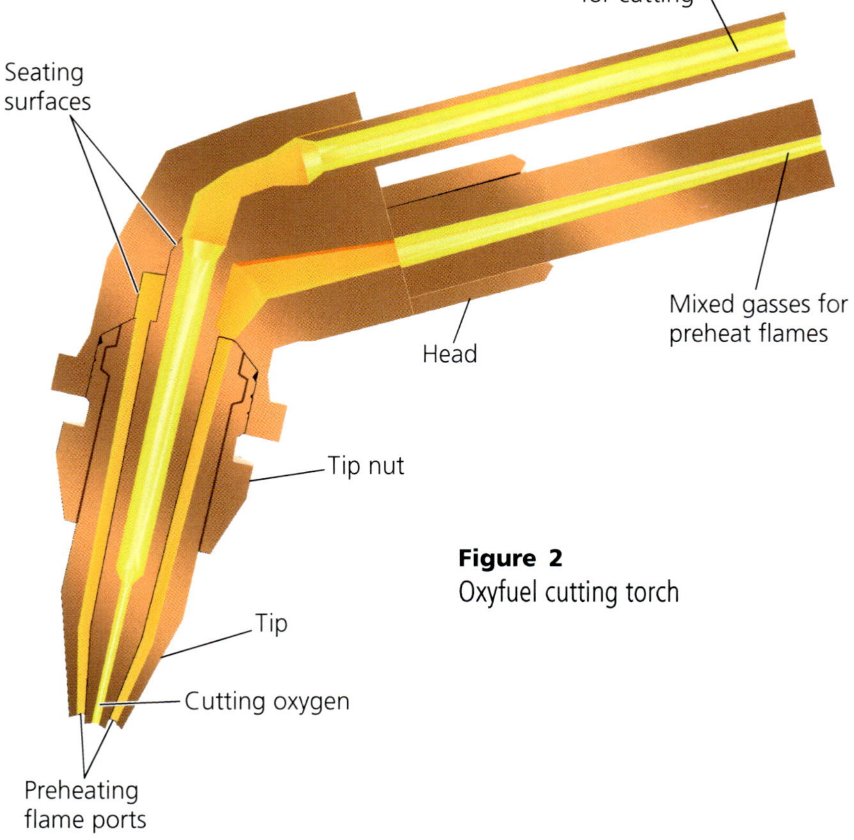

Figure 2
Oxyfuel cutting torch

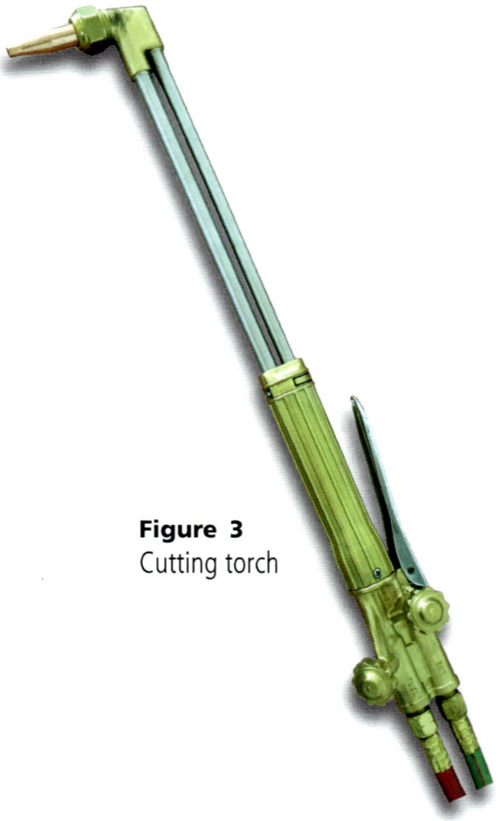

Figure 3
Cutting torch

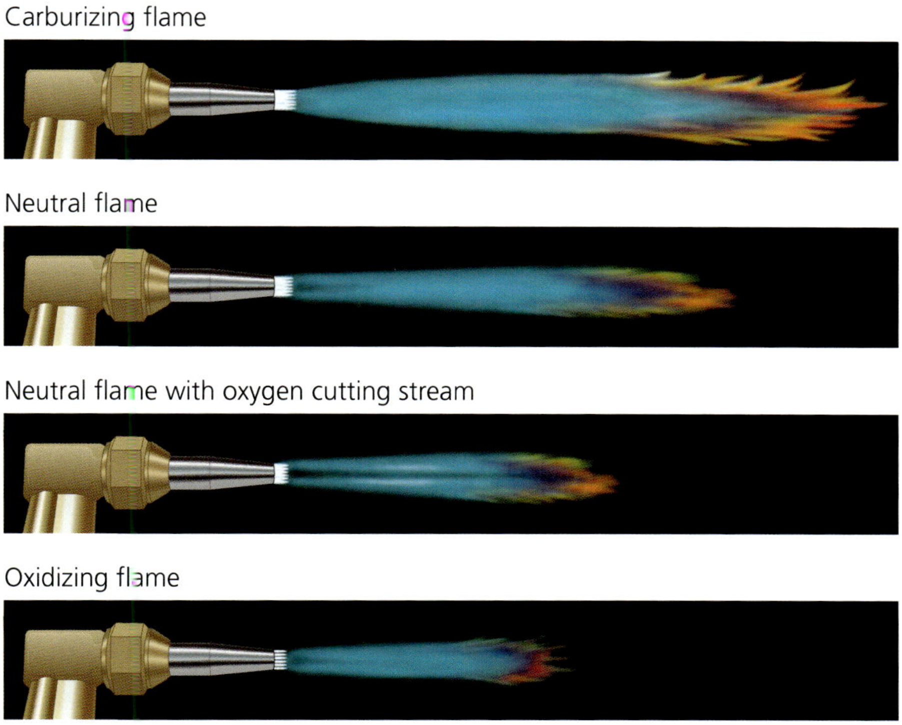

Figure 4
Oxyfuel cutting flames

- The fuel connectors use a similar setup. However, instead of a right-hand thread pattern, fuel connectors have a left-hand thread pattern and the nuts will be matched.

See Figure 5 for an illustration of the left-hand grooved nut pattern and the right-hand smooth nut pattern on standard oxygen and acetylene hose connections.

Check Valves and Flashback Arrestors A condition that can affect an oxyfuel torch is gas flowing in reverse. The *check valve* and *flashback arrestor* are special valves that prevent this from occurring. Check valves and flashback arrestors are similar. The check valve is used to prevent reverse flow into the hoses. The reverse flow is stopped by the check value before it can enter the hose.

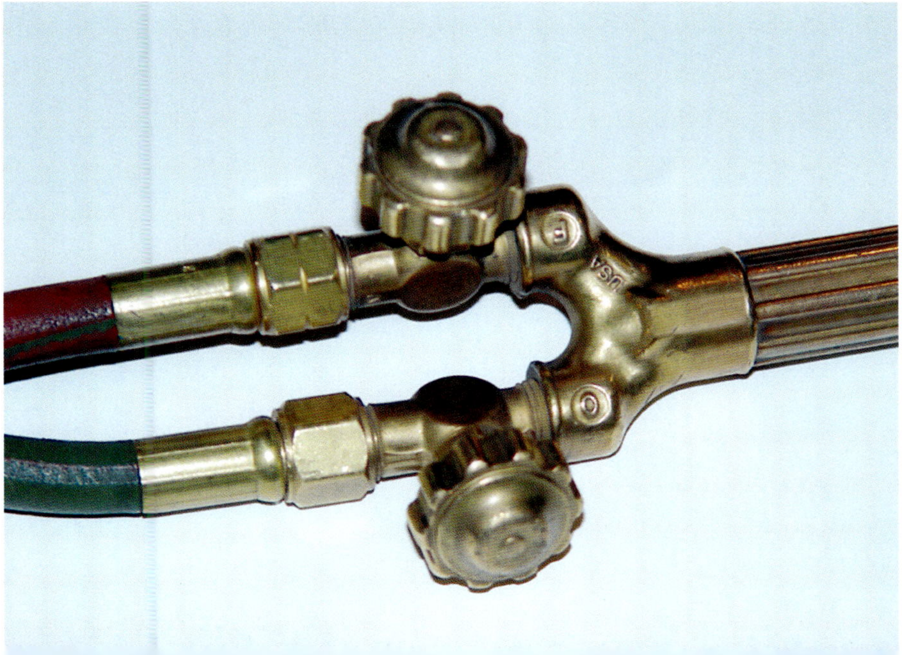

Figure 5
Hose connections

Check valves allow gas to flow in one direction only. See Figure 6. A flashback arrestor is a device installed at the hose/barrel connection point to stop a flashback from reaching the regulator or gas cylinder. The flashback is stopped at the flame block. It is important to use a flashback arrestor to prevent this from occurring. See Figure 7.

Torch valves are used to adjust the amount of oxygen and fuel gas mixture. Figure 8 shows a standard oxyfuel cutting torch.

There are two valves and one lever on the standard oxyfuel cutting torch. The oxygen valve adjusts the amount of oxygen in the torch and the lever controls the cutting pressure.

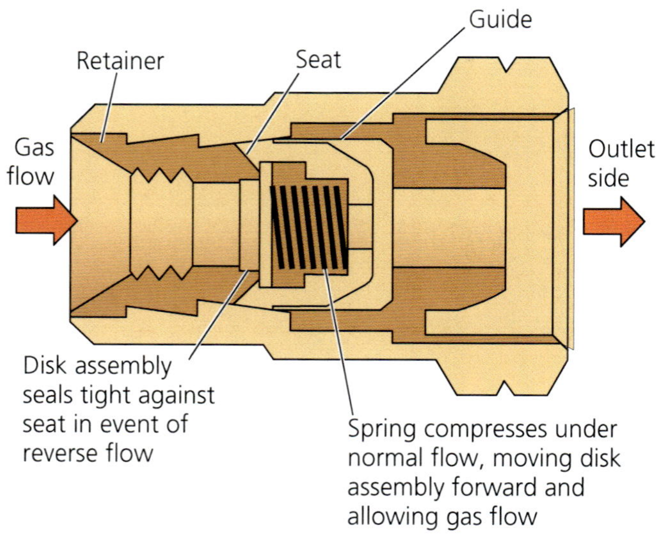

Figure 6
Check valve for an oxyfuel system

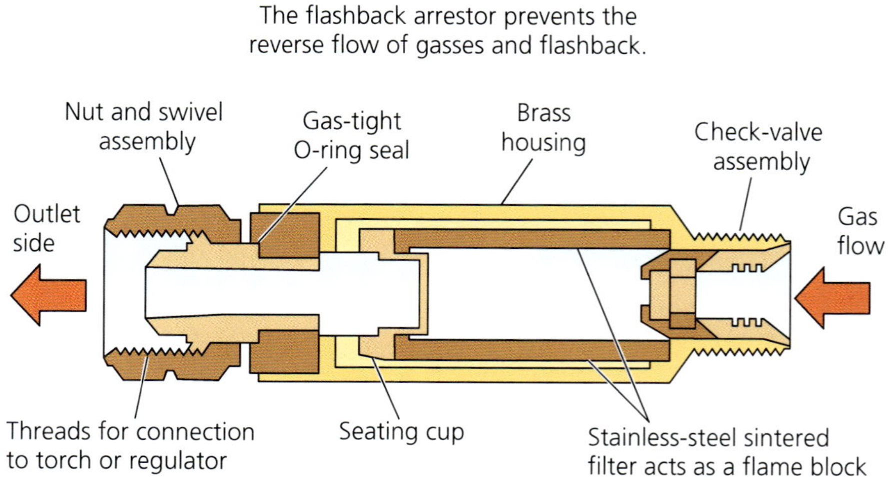

Figure 7
Flashback arrestor for an oxyfuel system

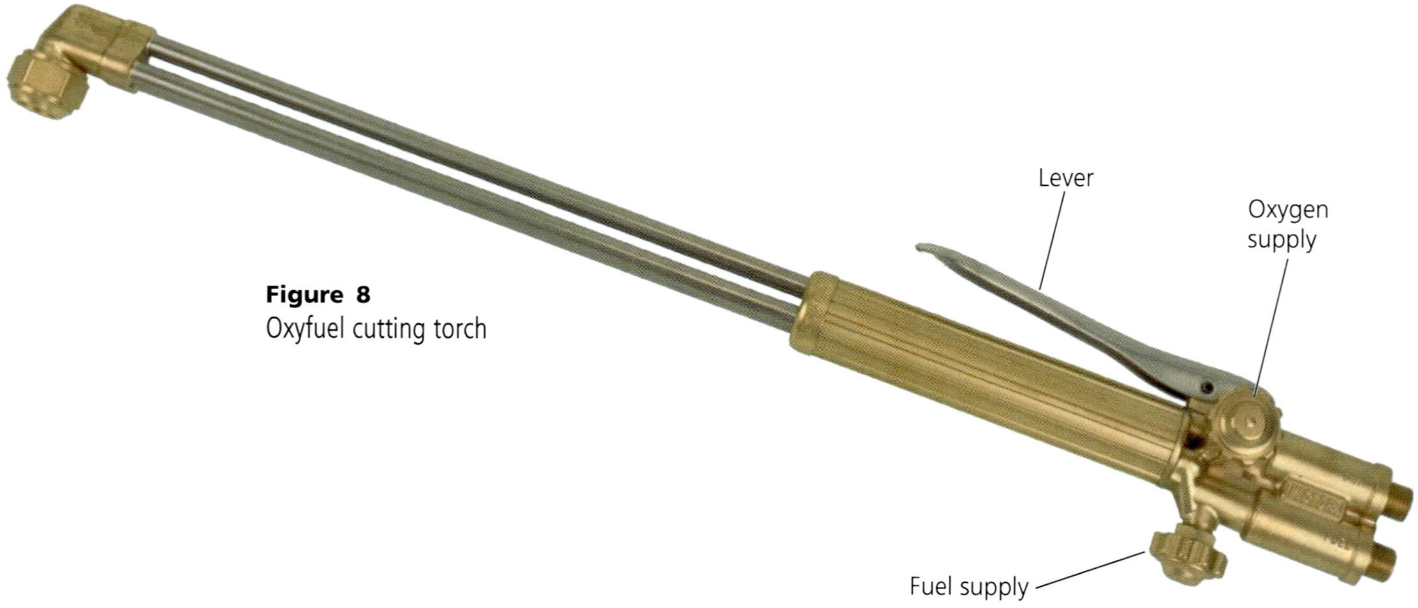

Figure 8
Oxyfuel cutting torch

PROCEDURE

How to Start Up an Oxyfuel System and Achieve a Neutral Flame

1. Be sure to wear the appropriate PPE for cutting: face shield with #5 shading, gloves, jacket, beanie, and safety glasses.
2. Make sure cylinders are secured to a wall or cart with a heavy-gauge chain before unscrewing and removing the protective caps on the gas cylinders.
3. Check the inside of the valves on the cylinders for debris, and then crack the cylinder valves to blow out any remaining debris.
4. Make sure the threads on the cylinders and gauges are undamaged and oil free.
5. Attach regulators to the cylinders. The gas cylinder has a left-handed and notched nut; the oxygen cylinder has a right-handed nut. Use an adjustable wrench or a cylinder wrench to tighten the regulators to the cylinders. Never use pliers and do not overtighten.
6. Make sure torch valves are closed.
7. Make sure the T-shaped handles on the cylinder gauges are loose enough to release the tension spring. The T-shaped handles should be loose to the touch.
8. Stand to the side of the fuel gauge and open the cylinder very slowly until the gas cylinder gauge creeps up. Open the cylinder valve only 1 ½ turns.
9. Turn the T-shaped handle on the gauge clockwise until the working gauge reads the proper pounds-per-square-inch (psi) for the tip being used. Never exceed 15 psi when using acetylene as fuel because it becomes unstable.
10. Stand to the side of the oxygen gauge and slowly open the cylinder valve until the oxygen cylinder gauge creeps up. Open the cylinder valve completely.
11. Turn the T-shaped handle on the gauge clockwise until the working gauge reads the proper psi for the tip being used.
12. Check for leaks at all connections with soapy water. Spray down each hose, torch, and cylinder connection.
13. Open the fuel valve ½ turn on the torch. Light the torch; only use a striker held about 1" away from the tip of the torch.
14. Once the torch is lit, open the fuel valve until black smoke clears up.
15. Open the oxygen valve on the torch slowly. The flame will slowly change color from yellow to blue and the inner flame will become very clear and focused. Compress the oxygen lever on the torch. The ¼" blue flame at the torch tip should not change color or become fuzzy. There will be three parts to the flame:
    - a brilliant inner core
    - a secondary cone on the edge of the inner core
    - a bluish outer envelope

PROCEDURE

How to Shut Down an Oxyfuel System

1. Shut the oxygen off on the torch handle first.
2. Next, shut the fuel valve off on the torch handle. If the torch is still burning, this indicates a fuel leak and the student should inform the teacher.
3. To bleed down the system, first shut off both cylinder valves completely.
4. Open the torch valves to relieve the gas pressure on the fuel and oxygen gauges. The pressure on the gauges will drop.
5. Turn the T-shaped handles on the gauges counterclockwise until the tension spring releases and the T-shaped handle is loose to the touch.
6. Use an adjustable wrench or a cylinder wrench to loosen the gauges from the cylinders. Take the gauges off the cylinders and make sure the valves on the torch are shut off.
7. Replace the protective cylinder caps.
8. Take the hoses off the gauges and torch.
9. Store all components in a clean, dry place.

Self Check

1. Which hose has a left-hand thread pattern?
2. What is backflow?
3. What changes the flame size on a torch?

Torch Tips

Most tips are made of pure drawn copper because this metal rapidly loses heat. Since copper is a soft metal and the torch tips undergo much use, they must be treated with care.

Tip Selection There are a variety of tips available. The choice of tip should be based on the following qualifications:

- the type of fuel gas being used
- the type of cutting that needs to be performed
- the thickness of the metal to be cut

Consult the manufacturer's guide for the torch that is being used. See Table 2 for an example of a manufacturer's guideline to tip selection.

TABLE 2 ***Tip and pressure selection for oxyfuel cutting***

Thickness of metal in inches	Tip size number	Cutting orifice drill size	Cutting speed (I.P.M.)	Oxygen pressure (P.S.I.G.)	Acetylene pressure (P.S.I.G.)	Methylactylene-propadiene (MAPP)® (P.S.I.G.)	Propane (P.S.I.G.)	Propylene (P.S.I.G.)	Natural gas (methane) (P.S.I.G.)
$^{1}/_{8}$″	000	72	20–30	$^{20}/_{25}$	$^{3}/_{5}$	10	10	$^{2}/_{5}$	10
$^{1}/_{4}$″	00	68	20–28	$^{20}/_{25}$	$^{3}/_{5}$	10	10	$^{2}/_{5}$	10
$^{3}/_{8}$″–$^{1}/_{2}$″	0	61	16–26	$^{25}/_{35}$	$^{3}/_{5}$	10	10	$^{3}/_{5}$	10
$^{3}/_{4}$″	1	57	15–20	$^{30}/_{35}$	$^{3}/_{5}$	10	10	$^{3}/_{6}$	10
1″–1 $^{1}/_{2}$″	2	54	13–18	$^{35}/_{40}$	$^{3}/_{6}$	10	10	$^{4}/_{8}$	10
2″–2 $^{1}/_{2}$″	3	51	10–12	$^{45}/_{50}$	$^{4}/_{8}$	10	10	$^{5}/_{9}$	10
3″	4	46	10–12	$^{40}/_{50}$	$^{5}/_{11}$	10	12	$^{6}/_{9}$	10

Keeping Tips Clean Torch tips are subject to wear, tear, and splatter. It is important to keep the torch tip clean and in good working order. A clean tip is vital for the following results:

- making a clean cut
- using less gas
- being more productive

To get tips completely clean, use a tip cleaner. See Figure 9. Tip cleaners clean the outside as well as the delicate inside of the torch tip. The tip should be removed before cleaning to ensure particles fall free from the torch. Use the correct sized tip cleaner. A tip cleaner that is too large could accidentally increase the size of the fuel holes, and therefore change the effectiveness of the tip. A tip cleaner that is too small won't completely clean the openings.

A flat, smooth tip with clean fuel and oxygen holes will produce a flame that is even across the cutting tip. A smooth file may be used to ensure the tip's end is flat. When using a file, make sure the tip is filed perpendicular to the flow of the gasses.

Figure 9
Tip cleaner

Lighting the Torch

safety tip

Use only a friction lighter to light an oxyfuel torch.

Flame The flame is a critical part of the welding process. Understanding how to start, control, and use flames can mean the difference between a mediocre cut and a good cut.

Oxyfuel Procedure The procedure below shows the proper start up and shutdown of a oxyfuel torch system. The oxyfuel torch should always be lit with a friction lighter. See Figure 10. Matches or butane lighters should never be used because these put the welder's hands too close to the lit flame. Besides this risk, a pack of matches or a butane lighter carried in a pocket might catch fire or explode during welding or cutting, causing serious burns.

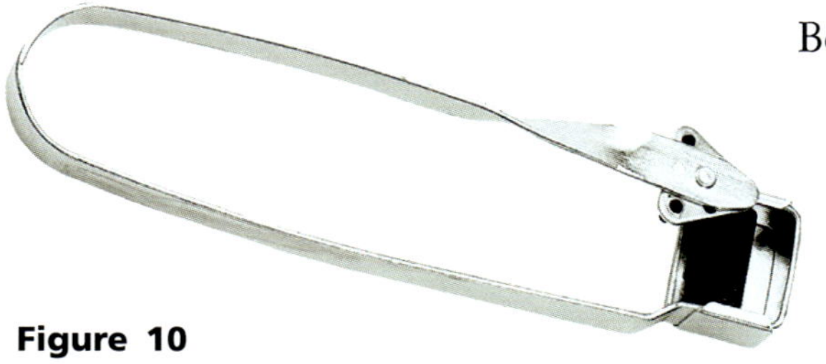

Figure 10
Triple-flint torch friction lighter

MAPP® gas, propane, and natural gas fuel torches are more difficult to light than acetylene. For these types of fuels, lay the tip of the torch against a metal plate to light with a friction lighter. The metal plate serves to concentrate the sparks and flame.

Types and Sizes of Flame Flame sizes and types vary according to the tip size, the type of fuel gas that is being used, and the fuel-to-oxygen ratio in the tip. Figure 11 shows the difference in Btu concentration between the acetylene flame and the flame of other gasses.

Flames can be adjusted to perform different cutting functions. These adjustments are made by varying the oxygen-to-fuel ratio. Figure 4, earlier in this chapter, shows the different kinds of oxyfuel flames and how they are used.

productivity tip

Do not set fuel pressure too high. Using too much fuel and oxygen does not produce a quicker cut but wastes gasses and empties the cylinders faster.

There are three basic types of cutting flames: neutral, carburizing, and oxidizing.

1. The *neutral flame* is the correct flame for efficient cutting. This flame is approximately 5,100° F to 5,600° F, depending on fuel.
2. The *carburizing flame* contains an excess of fuel. It is not as hot as the neutral flame.
3. The *oxidizing flame* contains an excess of oxygen. While it reduces preheating time, it is used only for rough work.

Flame Temperature Flame temperatures vary according to the fuel gas being used. Table 3 shows the gas efficiencies for various fuel gasses.

TABLE 3 ***Various fuel gas efficiencies***

Fuel gas	Btus (ft^3)	Combined intensity (suitable heat) Btu (s/ft^2 of flame cone area)	Flame temperature	Oxygen per ft^3 of fuel	Approx. normal velocity (ft/s)
Acetylene	1,433	12,700	5,420° F	1.04	17.7
MAPP®	2,381	5,540	5,301° F	2.4	7.9
Propane	2,309	5,500	5,190° F	4.00	11.9
Natural gas	918	5,600	5,000° F	1.50	15.2
Hydrogen	275	7,500	4,600° F	0.25	36

Flame temperatures can be adjusted higher or lower for any oxyfuel combination by adjusting the fuel-to-oxygen ratio. An oxidizing flame, that is, a flame that contains a higher proportion of oxygen than fuel gas, burns hotter than a neutral flame. This is the case no matter what fuel gas is used.

An experienced welder can tell by looking at the shape and color of the flame what the approximate temperature of the flame is. See Figure 4 to view different kinds of flames for an oxyfuel flame.

Figure 11
Btu concentration of the oxyfuel flame

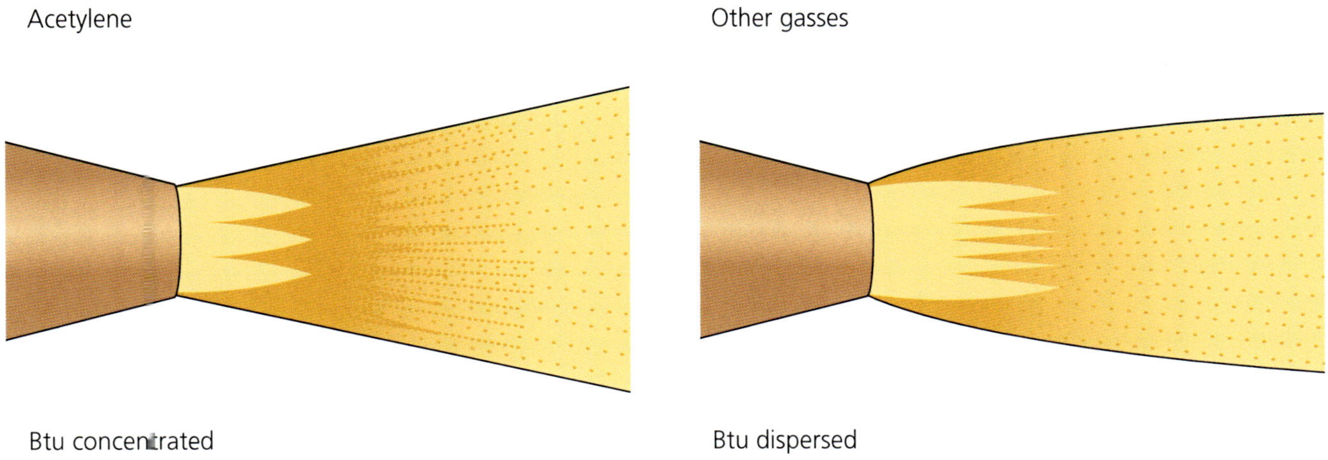

5 Cutting With Oxyfuel

When metal cutting needs to be done, there are several factors the welder must consider.

- What is the metal to be cut?
- How thick is the metal?
- What is the best fuel gas for this job, based on the previous two answsers?

Once these questions are answered and the appropriate fuel gas is chosen, it is time to begin cutting.

Before the Cut

The first thing the welder needs to do is to prepare the metal to be cut by preheating it.

Preheating the Metal Preheating is accomplished by lighting the torch, achieving a neutral flame, and directing the heat to the initial cutting location. Once the ignition temperature of the metal is reached, usually when it turns a bright cherry red color, it is time to apply the oxygen for cutting by depressing the oxygen lever.

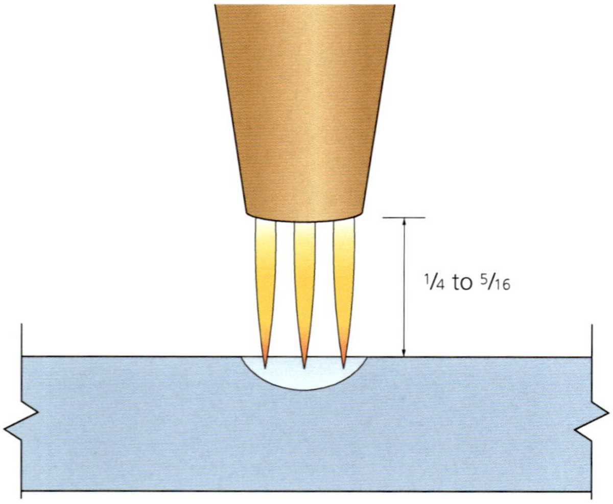

Figure 12
Correct torch position for cutting, end view

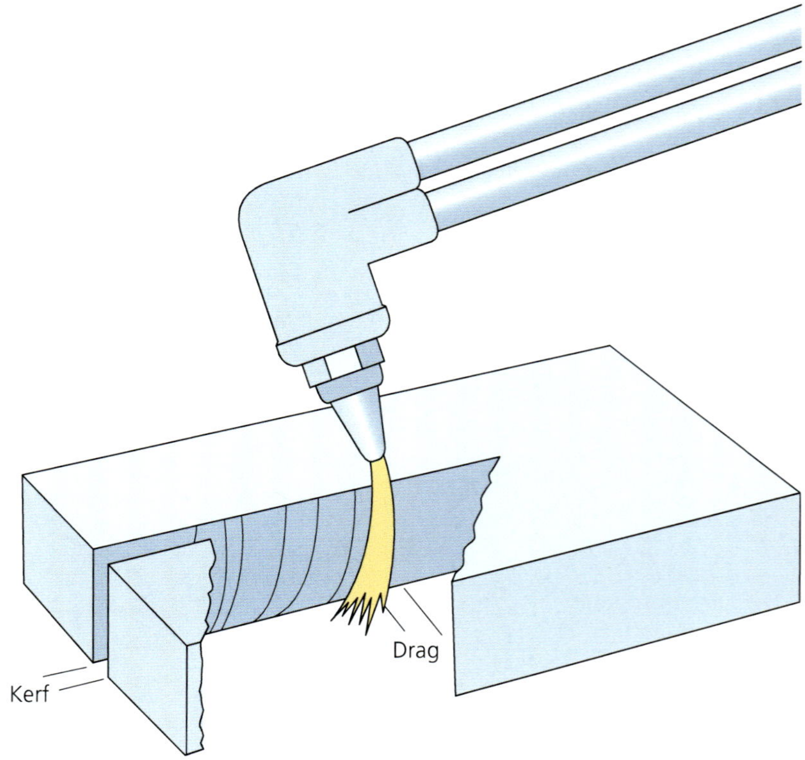

Figure 13
Correct torch position for cutting, side view

Positioning the Torch It is important to hold the torch in the correct position. See Figures 12 and 13 for the correct torch position. Remember, the fuel gas flame heats the metal while the oxygen stream does the cutting. The flame must travel first along the area to be cut, followed by the oxygen stream.

Cutting Terminology

Cutting can involve many angles and types of cuts. These are terms that a welder needs to know to describe cuts and their characteristics.

Kerf The **kerf** is the width of the cut.

Draglines *Draglines* are lines that appear on the oxygen-cut surface. Their contours and directions do not affect the quality of the cut surface.

Roughness The term *roughness* is used to mean the recurring peaks and valleys in the oxygen-cut surface. This can be determined by samples of acceptable quality or by comparison to the AWS C4.1 Surface Roughness Guide for Oxygen Cutting.

Top Edge Rounding When the welder melts the top edge of an oxygen-cut surface it is known as *top edge rounding.*

Notch A *notch* is a gouge in an oxygen-cut surface that is significantly deeper than the overall surface roughness.

Troubleshooting Oxyfuel Cutting

Figure 14 shows examples of oxyfuel cuts. Their defects are shown.

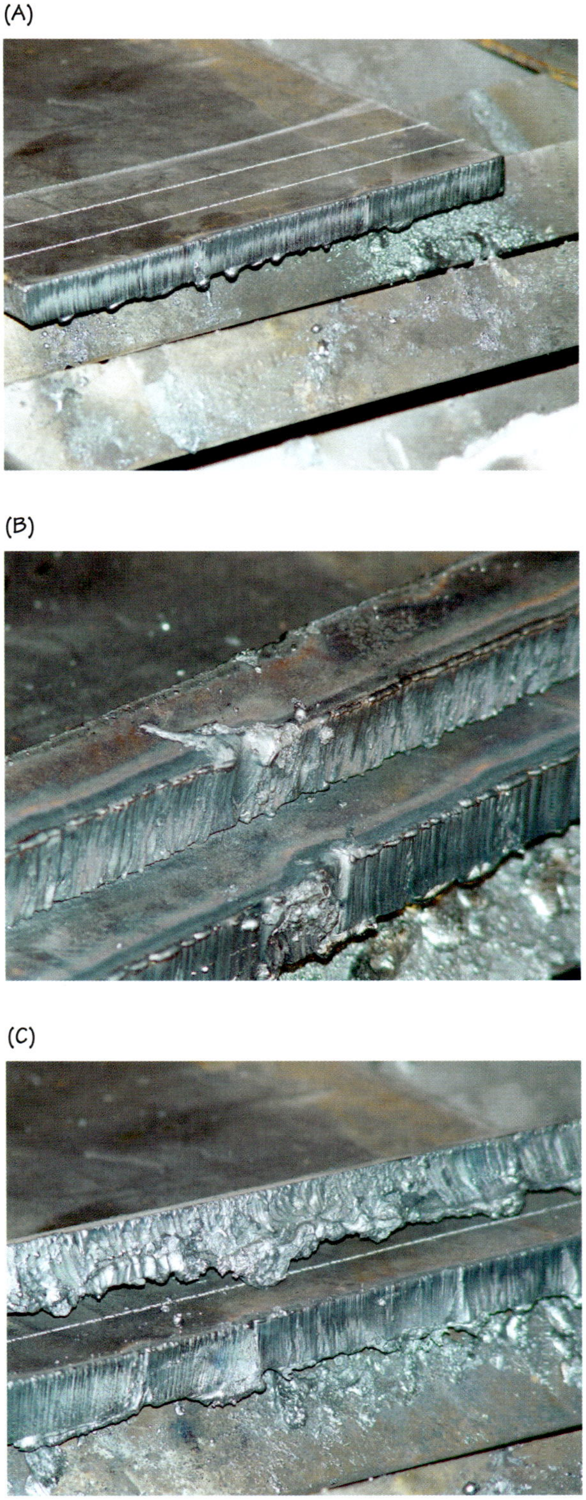

Figure 14
(A) Proper mixture creates a good cut
(B) Too much oxygen in mixture
(C) Too much acetylene in mixture

» The kerf is the width of the cut «

Self Check

1. Why is it important to use the correct-sized tip cleaner?
2. What can happen if the flame is not hot enough?
3. What happens if a cut is made too slowly?

Summary

Oxyfuel cutting is used to cut straight shapes and to scrap obsolete ferrous metal. A variety of fuel gasses is available. Choosing the right one to use depends on the type of metal to be cut. Acetylene gas is the most widely used fuel gas in oxyfuel cutting.

Chapter Review

Show your understanding of the information in this chapter by answering the questions and filling in the blanks below.

1. Oxyfuel cutting is used on ____________________ metals.

2. Which fuel gas has the lowest flash point?
   a. acetylene
   b. MAPP®
   c. propane
   d. natural gas

3. Which fuel gas is preferred for underwater cutting?
   a. acetylene
   b. MAPP®
   c. propane
   d. natural gas

4. The oxygen hose end has a left-handed thread. (True; False)

5. A(n) ____________________ ____________________ should be used to light a torch.

6. Under certain circumstances matches can be used to light MAPP® gas. (True; False)

7. What flame type most efficiently cuts steel? ____________________

8. An oxidizing flame can only be used for ____________________ cutting.

Project 1

Cutting Strips

Job Description

To complete this job successfully, the student should be able to lay out the project, set up and use an oxyfuel torch, and complete the project according to the given procedures. The cuts should be straight and have limited dross on the backside of the cut, with minimal variation and defects.

Materials

- one piece of ¼″ x 6″ bar stock

Tools

- wire brush
- chipping hammer
- #11 Vice Grips
- framing or speed square
- soap stone

Equipment

- oxyfuel torch set
- grinder with a wire wheel and a grinding wheel

Safety Precautions

- Use the grinder in a safe manner, without injury to self or others.
- Ensure the equipment is in proper working order
- Wear clear safety glasses at all times while in the shop.
- Wear appropriate clothing and work shoes at all times while in the shop.
- Wear the appropriate PPE while cutting.
- Keep the work area clean and safe.

Procedure

1. Measure an 8″ piece of ¼″ x 6″ bar stock.
2. Use the square to mark a line across the material at the 8″ mark.
3. To avoid stopping and starting more than necessary, sit or stand in a position that allows a complete range of motion for the work piece.
4. Without lighting the torch, practice the motion needed to complete the cut as described in the next step.
5. The torch can be pushed or pulled. If the torch is pushed, the operator must look over the torch head to see the marking to be cut. If the torch is pulled, the operator will be able to see the marking to be cut.
6. Use the oxyfuel torch with a #00 tip and cut the piece off the bar stock piece at the marked place.
7. Keep the cutting tip perpendicular to the plate, the blue tips just touching the plate.

Procedure, *cont.*

8. Measure five equally spaced lines across the 6″ direction of the plate. Stop the lines 1″ short of the end. This will leave the strips connected when the cutting is complete.
9. Remember to consider the kerf of the cut when measuring.
10. Use the square to draw straight lines.
11. Have the teacher check your layout before you proceed.
12. To avoid stopping and starting more than necessary, sit or stand in a position that allows the complete range of motion for the work piece.
13. Without lighting the torch, practice the motion needed to complete the cut as described in the next step.
14. The torch can be pushed or pulled. If the torch is pushed, the operator must look over the torch head to see the markings to be cut. If the torch is pulled, the operator will be able to see the markings to be cut.
15. Use the oxyfuel torch with a #00-tip and cut the lines, starting at the 1″ mark.
16. Dip the piece in the cooling tank.
17. When cleaning the piece, hold it with the #11 vise grips.
18. Clean the dross off of the plate with a chipping hammer.
19. Let the teacher see the finished product for grading.
20. Save the plate to use for welding practice later; it is great practice for the E60XX and E7018 electrodes.

Cutting Strips Project Rubric

POINTS:	5	4	3	2	1	0
Criteria:						
Overall dimension	± ⅛″	± ¼″	± ⅜″	± ½″	± ⅝″	± ¾″
Layout	± ⅛″	± ¼″	± ⅜″	± ½″	± ⅝″	± ¾″
Fuel setting	± 0 psi	± 1 psi	± 2 psi	± 3 psi	± 4 psi	± 5 psi
Oxygen setting	± 5 psi	± 10 psi	± 15 psi	± 20 psi	± 25 psi	± 30 psi
Correct tip Size	#00 tip used	± one tip size	± two tip sizes			± three tip sizes
Straight lines	± ⅛″	± ¼″	± ⅜″	± ½″	± ⅝″	± ¾″
Restarts	0 to 1	2 to 3	4 to 5	6 to 7	8 to 9	10 to 11
Kerf (not cut or heavy dross)	0 to 2″	2″ to 4″	4″ to 6″	6″ to 8″	8″ to 10″	+ 10″

Student Name: ______________________________

Date: ______________________________

Cutting Strips Project Evaluation

PROCEDURE	CRITERIA	POSSIBLE POINTS	SCORE
Layout	Overall dimension correct 6″x 8″	5	
	Layout of project correct	5	
	Subtotal	*10*	
Oxyfuel Set Up	Fuel setting correct	5	
	Oxygen setting correct	5	
	Correct tip size	5	
	Subtotal	*15*	
Cuts	Straight lines	5	
	Restarts	5	
	Kerf	5	
	Subtotal	*15*	
General	Tool handling	5	
	Followed directions	5	
	Safe work practices	5	
	Cleaned up	5	
	Subtotal	*20*	
		Student score:	

Total Possible Points = 60

Suggested minimum acceptable score: 42 points or _______

Student's Signature: ______________________________

Teacher's Signature: ______________________________

Student Name:______________________________
Date:______________________________

Cutting Strips Project Safety Evaluation

Competency	Demonstrated			Date Completed	Teacher Initial*	Student Initial**	Comments
	Yes	No	N/A				
1. Wore proper safety equipment							
2. Always worked in a clean work area							
3. Used tools only as designed							
4. Always set up torch properly							
5. Lit torch safely							
6. Handled gas tanks properly							
7. Aware of spark direction while grinding							

*Teacher's initial indicates that the student demonstrated competencies for proper and safe use of this tool.
**Student's initial indicates that the student demonstrated competencies for proper and safe use of this tool.

Student's Signature: ______________________________

Teacher's Signature: ______________________________

Project 2

Cutting T-Joints

Job Description

To complete this job successfully, the student should be able to lay out the project, set up and use an oxyfuel torch, and complete the project according to the given procedures. The cuts should be straight and have limited dross on the backside of the cut, with minimal variation and defects.

Materials

- one piece of 3/8″ x 3″ bar stock

Tools

- wire brush
- chipping hammer
- #11 Vice Grips
- framing or speed square
- soap stone

Equipment

- oxyfuel torch set
- grinder with a wire wheel and a grinding wheel

Safety Precautions

- Use the grinder in a safe manner, without injury to self or others.
- Ensure the equipment is in proper working order
- Wear clear safety glasses at all times while in the shop.
- Wear appropriate clothing and work shoes at all times while in the shop.
- Wear the appropriate PPE while cutting.
- Keep the work area clean and safe.

Procedure

1. Measure a 3″ piece of 3/8″ x 7″ bar stock.
2. Using the square, mark a line across the material at the 3″ mark.
3. To avoid stopping and starting more than necessary, sit or stand in a position that allows a complete range of motion for the work piece.
4. Without lighting the torch, practice the motion needed to complete the cut as described in the next step.
5. The torch can be pushed or pulled. If the torch is pushed, the operator must look over the torch head to see the marking to be cut. If the torch is pulled, the operator will be able to see the marking to be cut.
6. Using the oxyfuel torch with a #0 tip, cut the piece off the bar stock piece.

Procedure, *cont.*

7. Keep the cutting tip perpendicular to the plate, the blue tips just touching the plate.
8. Measure in 1 ½″ across the 3″ side of the piece.
9. Remember to consider the kerf of the cut when measuring.
10. Using a square, make a straight line down the 1 ½″ mark.
11. Let the teacher check the layout before you proceed.
12. To avoid stopping and starting more than necessary, sit or stand in a position that allows the complete range of motion for the work piece.
13. Without lighting the torch, practice the motion it will take to complete the cut as described in the next step.
14. The torch can be pushed or pulled. If the torch is pushed, the operator must look over the torch head to see the markings to be cut. If the torch is pulled, the operator will be able to see the markings to be cut.
15. Using the oxyfuel torch with a #0 tip, bevel both 7″ sides of the piece at a 45° angle.
16. Cut the line marked at 1 ½″.
17. Dip the piece in the cooling tank.
18. When cleaning the piece, hold it with the #11 vise grips.
19. Clean the dross off of the plate with a chipping hammer.
20. Let the teacher see the finished product for grading.
21. Save the pieces for welding practice later.

Cutting T-Joints Rubric

POINTS:	5	4	3	2	1	0
Criteria:						
Overall dimension	± ⅛″	± ¼″	± ⅜″	± ½″	± ⅝″	± ¾″
Layout	± ⅛″	± ¼″	± ⅜″	± ½″	± ⅝″	± ¾″
Fuel setting	± 0 psi	± 1 psi	± 2 psi	± 3 psi	± 4 psi	± 5 psi
Oxygen setting	± 5 psi	± 10 psi	± 15 psi	± 20 psi	± 25 psi	± 30 psi
Correct tip Size	#0 tip used	± one tip size	± two tip sizes			± three tip sizes
Straight lines	± ⅛″	± ¼″	± ⅜″	± ½″	± ⅝″	± ¾″
Restarts	0 to 1	2 to 3	4 to 5	6 to 7	8 to 9	10 to 11
Kerf (not cut or heavy dross)	0 to 1″	1″ to 2″	2″ to 3″	3″ to 4″	4″ to 5″	+ 5″
Bevels	± 0	± 5°	± 10°	± 15°	± 20°	± 25°

Student Name: ____________________

Date: ____________________

Cutting T-Joints Evaluation

PROCEDURE	CRITERIA	POSSIBLE POINTS	SCORE
Layout	Overall dimension correct 3″x 7″	5	
	Layout of project correct	5	
	Subtotal	*10*	
Oxyfuel Setup	Fuel setting correct	5	
	Oxygen setting correct	5	
	Correct tip size	5	
	Subtotal	*15*	
Cuts	Straight lines	5	
	Restarts	5	
	Kerf	5	
	Bevels	5	
	Subtotal	*20*	
General	Tool handling	5	
	Followed directions	5	
	Safe work practices	5	
	Cleaned up	5	
	Subtotal	*20*	
		Student score:	

Total Possible Points = 65

Suggested minimum acceptable score: 46 points or ______

Student's Signature: ____________________

Teacher's Signature: ____________________

Student Name: ____________________

Date: ____________________

Cutting T-Joints Safety Evaluation

Competency	Demonstrated			Date Completed	Teacher Initial*	Student Initial**	Comments
	Yes	No	N/A				
1. Wore proper safety equipment							
2. Always worked in a clean work area							
3. Used tools only as designed							
4. Always set up torch properly							
5. Lit torch safely							
6. Handled gas tanks properly							
7. Aware of spark direction while grinding							

*Teacher's initial indicates that the student demonstrated competencies for proper and safe use of this tool.

**Student's initial indicates that the student demonstrated competencies for proper and safe use of this tool.

Student's Signature: ____________________

Teacher's Signature: ____________________

Project 3

Cutting Lap Joints

Job Description

To complete this job successfully, the student should be able to lay out the project, set up and use an oxyfuel torch, and complete the project according to the given procedures. The cuts should be straight and have limited dross on the backside of the cut, with minimal variation and defects.

Materials

- one piece of ¼″ x 6″ bar stock

Tools

- wire brush
- chipping hammer
- #11 Vice Grips
- framing or speed square
- soap stone

Equipment

- oxyfuel torch set
- grinder with a wire wheel and a grinding wheel

Safety Precautions

- Use the grinder in a safe manner, without injury to self or others.
- Ensure the equipment is in proper working order
- Wear clear safety glasses at all times while in the shop.
- Wear appropriate clothing and work shoes at all times while in the shop.
- Wear the appropriate PPE while cutting.
- Keep the work area clean and safe.

Procedure

1. Measure a 3″ piece of ¼″ x 6″ bar stock.
2. Using the square, mark a line across the material at the 3″ mark.
3. To avoid stopping and starting more than necessary, sit or stand in a position that allows a complete range of motion for the work piece.
4. Without lighting the torch, practice the motion needed to complete the cut as described in the next step.
5. The torch can be pushed or pulled. If the torch is pushed, the operator must look over the torch head to see the marking to be cut. If the torch is pulled, the operator will be able to see the marking to be cut.
6. Using the oxyfuel torch with a #00 tip, cut the piece off the bar stock piece.
7. Keep the cutting tip perpendicular to the plate, the blue tips just touching the plate.

8. Measure in 1 ½″ across the 3″ side of the piece.
9. Remember to consider the kerf of the cut when measuring.
10. Using a square make a straight line down the 1 ½″ mark.
11. Let the teacher check the layout before you proceed.
12. To avoid stopping and starting more than necessary, sit or stand in a position that allows the complete range of motion for the work piece.
13. Without lighting the torch, practice the motion it will take to complete the cut.
14. The torch can be pushed or pulled. If the torch is pushed, the operator must look over the torch head to see the markings to be cut. If the torch is pulled, the operator will be able to see the markings to be cut.
15. Using the oxyfuel torch with a #00-tip, cut the line marked at 1 ½″.
16. Dip the piece in the cooling tank.
17. When cleaning the piece, hold the piece with the #11 vise grips.
18. Clean the dross off of the plate with a chipping hammer.
19. Let the teacher see the finished product for grading.
20. Save the pieces for welding practice later.

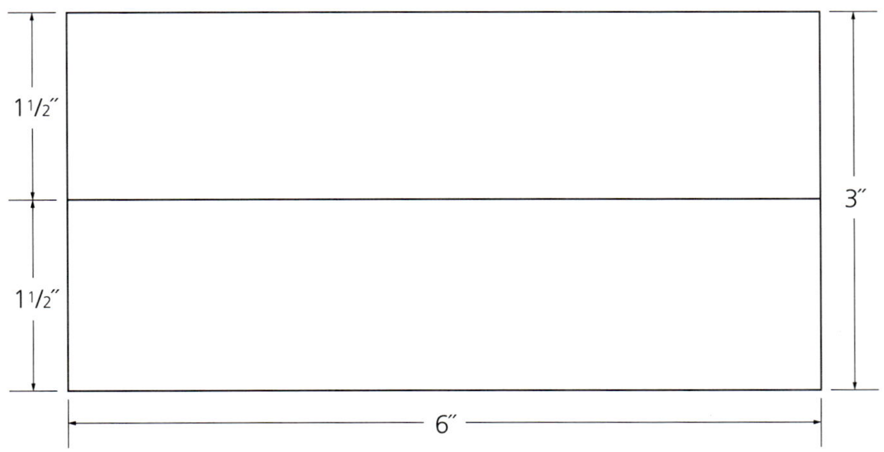

Cutting Lap Joints Rubric

POINTS:	5	4	3	2	1	0
Criteria:						
Overall dimension	± ⅛″	± ¼″	± ⅜″	± ½″	± ⅝″	± ¾″
Layout	± ⅛″	± ¼″	± ⅜″	± ½″	± ⅝″	± ¾″
Fuel setting	± 0 psi	± 1 psi	± 2 psi	± 3 psi	± 4 psi	± 5 psi
Oxygen setting	± 5 psi	± 10 psi	± 15 psi	± 20 psi	± 25 psi	± 30 psi
Correct tip Size	#00 tip used	± one tip size	± two tip sizes			± three tip sizes
Straight lines	± ⅛″	± ¼″	± ⅜″	± ½″	± ⅝″	± ¾″
Restarts	0 to 1	2 to 3	4 to 5	6 to 7	8 to 9	10 to 11
Kerf (not cut or heavy dross)	0 to 1″	1″ to 2″	2″ to 3″	3″ to 4″	4″ to 5″	+ 5″

Student Name: ____________________
Date: ____________________

Cutting Lap Joints Evaluation

PROCEDURE	CRITERIA	POSSIBLE POINTS	SCORE
Layout	Overall dimension correct 3″x 6″	5	
	Layout of project correct	5	
	Subtotal	*10*	
Oxyfuel Setup	Fuel setting correct	5	
	Oxygen setting correct	5	
	Correct tip size	5	
	Subtotal	*15*	
Cuts	Straight lines	5	
	Restarts	5	
	Kerf	5	
	Subtotal	*15*	
General	Tool handling	5	
	Followed directions	5	
	Safe work practices	5	
	Cleaned up	5	
	Subtotal	*20*	
		Student score:	

Total Possible Points = 60

Suggested minimum acceptable score: 42 points or _______

Student's Signature: ____________________

Teacher's Signature: ____________________

Student Name: ____________________

Date: ____________________

Cutting Lap Joints Safety Evaluation

Competency	Demonstrated			Date Completed	Teacher Initial*	Student Initial**	Comments
	Yes	No	N/A				
1. Wore proper safety equipment							
2. Always worked in a clean work area							
3. Used tools only as designed							
4. Always set up torch properly							
5. Lit torch safely							
6. Handled gas tanks properly							
7. Aware of spark direction while grinding							

*Teacher's initial indicates that the student demonstrated competencies for proper and safe use of this tool.

**Student's initial indicates that the student demonstrated competencies for proper and safe use of this tool.

Student's Signature: ____________________

Teacher's Signature: ____________________

Project 4

Cutting Shapes

Job Description

To complete this job successfully, the student should be able to lay out the project, set up and use an oxyfuel torch, and complete the project according to the given procedures. The cuts should be straight and have limited dross on the backside of the cut, with minimal variation and defects.

Materials

- one piece of ⅜″ x 7″ bar stock

Tools

- wire brush
- chipping hammer
- #11 Vice Grips
- framing or speed square
- compass
- soap stone

Equipment

- oxyfuel torch set
- grinder with a wire wheel and a grinding wheel

Safety Precautions

- Use the grinder in a safe manner, without injury to self or others.
- Ensure the equipment is in proper working order
- Wear clear safety glasses at all times while in the shop.
- Wear appropriate clothing and work shoes at all times while in the shop.
- Wear the appropriate PPE while welding.
- Keep the work area clean and safe.

Procedure

1. Measure a 12″ piece of ⅜″ x 7″ bar stock.
2. Using the square, mark a line across the material at the 12″ mark.
3. To avoid stopping and starting more than necessary, sit or stand in a position that allows a complete range of motion for the work piece.
4. Without lighting the torch, practice the motion needed to complete the cut as described in the next step.
5. The torch can be pushed or pulled. If the torch is pushed, the operator must look over the torch head to see the marking to be cut. If the torch is pulled, the operator will be able to see the marking to be cut.
6. Using the oxyfuel torch with a #0 tip, cut the piece off the bar stock piece.
7. Keep the cutting tip perpendicular to the plate, the blue tips just touching the plate.

8. Measure out and mark on the plate all of the given shapes shown in the diagram. Notice that all of the measurements are from one end and one side.
9. Remember to consider the kerf of the cut when measuring.
10. Use the square to draw the lines straight.
11. The teacher will check your layout before you proceed.
12. To avoid stopping and starting more than necessary, sit or stand in a position that allows the complete range of motion for the work piece.
13. Without lighting the torch, practice the motion it will take to complete the cut.
14. The torch can be pushed or pulled. If the torch is pushed, the operator must look over the torch head to see the markings to be cut. If the torch is pulled, the operator will be able to see the marking to be cut.
15. For circles, start in the center and circle into the cut to get the best transition onto the line.
16. Using the oxyfuel torch with a #0 tip, cut the lines.
17. Dip the piece in the cooling tank.
18. Hold the piece with the #11 vise grips for cleaning.
19. Clean the dross off of the plate with a chipping hammer.
20. Let the teacher see the finished product for grading.
21. Save the plate for welding practice later. It is a great piece to tack your welding projects to. This will save the surface of the welding table.

Project #4 Cutting Shapes

3/8″ x 7″ x 12″

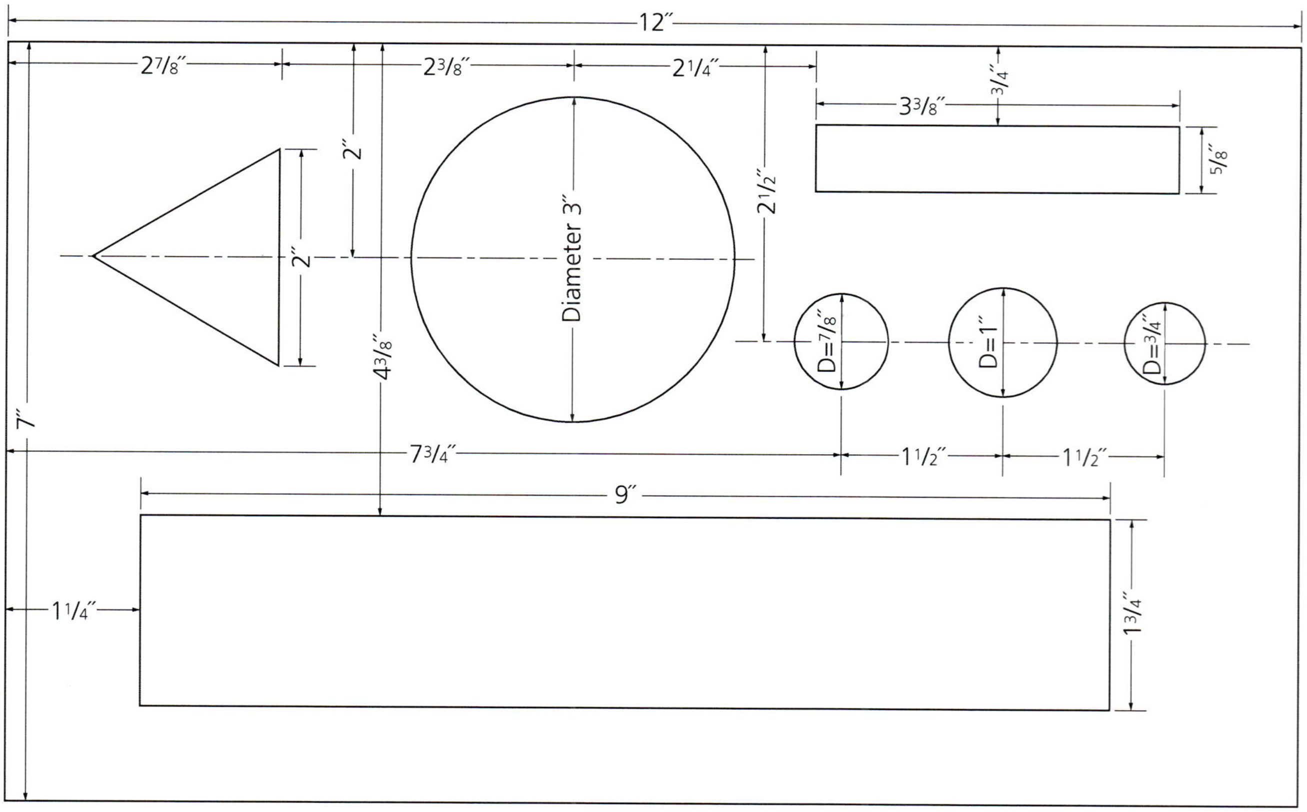

Cutting Shapes Rubric

POINTS:	5	4	3	2	1	0
Criteria:						
Overall dimension	± 1/8″	± 1/4″	± 3/8″	± 1/2″	± 5/8″	± 3/4″
Triangle layout	± 1/8″	± 1/4″	± 3/8″	± 1/2″	± 5/8″	± 3/4″
Circle layout	± 1/8″	± 1/4″	± 3/8″	± 1/2″	± 5/8″	± 3/4″
Small rectangle layout	± 1/8″	± 1/4″	± 3/8″	± 1/2″	± 5/8″	± 3/4″
Small circles layout	± 1/8″	± 1/4″	± 3/8″	± 1/2″	± 5/8″	± 3/4″
Large rectangle layout	± 1/8″	± 1/4″	± 3/8″	± 1/2″	± 5/8″	± 3/4″
Fuel setting	± 0 psi	± 1 psi	± 2 psi	± 3 psi	± 4 psi	± 5 psi
Oxygen setting	± 5 psi	± 10 psi	± 15 psi	± 20 psi	± 25 psi	± 30 psi
Correct tip Size	#0 tip used	± one tip size	± two tip sizes			± three tip sizes
Straight lines	± 1/8″	± 1/4″	± 3/8″	± 1/2″	± 5/8″	± 3/4″
Restarts	0 to 1	2 to 3	4 to 5	6 to 7	8 to 9	10 to 11
Triangle	± 1/8″	± 1/4″	± 3/8″	± 1/2″	± 5/8″	± 3/4″
Large circle	± 1/8″	± 1/4″	± 3/8″	± 1/2″	± 5/8″	± 3/4″
Small rectangle	± 1/8″	± 1/4″	± 3/8″	± 1/2″	± 5/8″	± 3/4″
Small circles	± 1/8″	± 1/4″	± 3/8″	± 1/2″	± 5/8″	± 3/4″
Large rectangle	± 1/8″	± 1/4″	± 3/8″	± 1/2″	± 5/8″	± 3/4″

Student Name:____

Date:____

Cutting Shapes Evaluation

PROCEDURE	CRITERIA	POSSIBLE POINTS	SCORE
Layout	Overall dimension correct 7″x 12″	5	
	Triangle	5	
	Circle	5	
	Small rectangle	5	
	Small circles	5	
	Large rectangle	5	
	Subtotal	*30*	
Oxyfuel Setup	Fuel setting correct	5	
	Oxygen setting correct	5	
	Correct tip size	5	
	Subtotal	*15*	
Cuts	Straight lines	5	
	Restarts	5	
	Triangle	5	
	Circle	5	
	Small rectangle	5	
	Small circles	5	
	Large rectangle	5	
	Subtotal	*35*	
General	Tool handling	5	
	Followed directions	5	
	Safe work practices	5	
	Cleaned up	5	
	Subtotal	*20*	
Total Possible Points = 100		**Student score:**	

Suggested minimum acceptable score: 70 points or ______

Student's Signature: ______________________________________

Teacher's Signature: ______________________________________

Student Name: ____

Date: ____

Cutting Shapes Safety Evaluation

Competency	Demonstrated			Date Completed	Teacher Initial*	Student Initial**	Comments
	Yes	No	N/A				
1. Wore proper safety equipment							
2. Always worked in a clean work area							
3. Used tools only as designed							
4. Always set up torch properly							
5. Lit torch safely							
6. Handled gas tanks properly							
7. Aware of spark direction while grinding							

*Teacher's initial indicates that the student demonstrated competencies for proper and safe use of this tool.

**Student's initial indicates that the student demonstrated competencies for proper and safe use of this tool.

Student's Signature: ____

Teacher's Signature: ____

CHAPTER 4

Welding Equipment

There are several types of shielded metal arc welding (SMAW) machines. These machines produce either alternating current (AC) or direct current (DC). It is important to understand how these machines work so that welding can be performed effectively. Two very important components of the welding machine are the electrode holder and the work clamp. The electrode holder is what the welder grips to feed the weld. The work clamps create a ground between the welding machine and the work. Clamps are connected to cables, or leads, which run from the welding machine to the work. They conduct current from the welding machine to the work and back. If a cable is damaged or not functioning properly, the result could be a bad weld or even electrocution.

Contents

Objectives

- Recognize and identify the different types of welding machines.
- Demonstrate an understanding of what an AC/DC transformer is and how it works.
- Demonstrate an understanding of what a rectifier and an inverter are.
- Demonstrate an understanding of the types of generators used in welding.
- Explain different types of electrode holders and their parts.
- Discuss and know how to inspect the work clamp, cables, and leads.
- Discuss the types of available welding tools and how these tools are used.

Key Terms

alternating current (AC) electric current that reverses direction at regular intervals

arc welding various welding processes that use an electric arc from a metal rod to the workpiece to create a weld

cables or **leads** insulated electrical conductor composed of thousands of flexible thin copper wires

chipping hammer double-ended tool to chip slag in corners, edges, and ripples, and for general cleaning

constant current when the current does not change based on workload

constant voltage when voltage does not change based on workload

direct current (DC) electric current that flows in a constant direction

duty cycle amount of time a welding machine can run without overheating

electrode holder clamping device used to hold the electrode for welding

inverter changes frequency to convert high-voltage, low-amperage AC current to low-voltage, high-amperage DC current

output slope change in relationship between the output voltage and the output current or amperage of the machine as the current or welding workload is changed

shielded metal arc welding (SMAW) or **stick welding** process that uses an electric current (AC or DC) to form an arc between an electrode coated in flux and the metals to be joined

wire brush removes rust, slag, and other dry materials from a workpiece

work clamp device used to produce a ground between the welding machine and the work

1 Types of Welding

Before discussing welding equipment, this section briefly outlines some different types of welding. In **arc welding**, welding takes place when metal is melted from an electrode and carried across the electric arc to deposit filler metal in a joint. There are several different types of arc welding performed today: shielded metal arc welding, called SMAW or stick welding, gas tungsten arc welding, often referred to as GTAW or TIG welding, gas metal arc welding, otherwise known as GMAW or MIG or MAG welding, and Flux Cored Arc Welding (FCAW).

Shielded Metal Arc Welding (SMAW)

A popular technique, **shielded metal arc welding** (**SMAW**) is commonly known as **stick welding**. This technique does not use pressure. Rather, the weld occurs when the heat from the electric arc brings both the metal to be welded and the electrode to a molten state. The heat melts the electrode and the base metal, that is the metal to be welded, near it. The droplets from the electrode gather in a molten pool in the base metal to create the weld.

SMAW is used on ferrous and some nonferrous metals. It is especially effective on the following metals and alloys.

- carbon steel
- low-alloy steel
- stainless steel
- cast iron
- nickels and their alloys
- aluminum and its alloys

Gas Tungsten Arc Welding (GTAW)

The techniques referred to as gas tungsten arc welding (GTAW), also referred to as tungsten inert gas welding (TIG welding), use a nonconsumable tungsten electrode to sustain the arc. The tungsten electrode is shielded from the atmosphere by an inert gas. GTAW is versatile in that a weld can be made with or without a filler metal.

In addition, metal less than one millimeter thick can be welded using this technique.

GTAW is used on ferrous and nonferrous metals or with difficult-to-weld metals and alloys such as aluminum, stainless steel, magnesium, and titanium.

Gas Metal Arc Welding (GMAW)

The initial purpose of gas metal arc welding (GMAW) was to weld nonferrous metals, specifically magnesium and aluminum, for the aircraft industry. GMAW is also known as metal inert gas welding (MIG welding) or metal active gas welding (MAG welding) depending on the gas used. In the field, it is frequently called wire feed welding.

GMAW is also useful on other metals and their alloys, including nickel, magnesium, titanium, copper, and steel.

Flux Cored Arc Welding (FCAW)

Flux cored arc welding (FCAW) is another type of wire feed welding. FCAW uses a continuous-feed consumable wire electrode whose core is filled with flux and alloying agents. There are two types of FCAW: gas-shielded (FCAW-G) and self-shielded (FCAW-S). In FCAW-G, argon and carbon dioxide are the two main gasses used to provide the additional shielding needed for this process. The electrodes used in FCAW-S provide their own shielding gas and slag upon melting and do not require an external shielding gas. FCAW can be used on ferrous and nonferrous metals.

Self Check

1. What does SMAW stand for?
2. Which process can be used with or without filler metal?

2 Power Sources

There are a number of power sources for welding equipment. These include AC-DC transformers, transformer-rectifiers, inverters, and generators. Although AC/DC transformer-rectifier sets are still used in the industry today, more modern technology is available. Even with this new technology, the same principles of welding apply.

AC Transformers, AC/DC Transformer-Rectifier Set

Certain metals and alloys require **alternating current** (**AC**) power to weld more efficiently, while **direct current** (**DC**) power is required for other metals. An AC-DC welding machine has both power types available in one machine. AC-DC welder machines are used for SMAW and GTAW.

To select between these two power outputs, the welder simply flips a switch or changes where the **cables**, or **leads**, are plugged into the machine. Figure 1 shows an AC-DC welding machine. These machines have a built-in rectifier to convert AC current to DC. This feature allows welders to choose the current best suitable for welding the metal on which they are working.

Figure 1
300-amp AC-DC GTAW/SMAW machine

DC Transformer-Rectifier

The DC transformer-rectifier is a versatile machine. It can be used for GTAW and SMAW on multiple-operator systems. It supports direct current electrode negative (DCEN) welding, also called straight polarity welding, or direct current electrode positive (DCEP) welding, also called reverse polarity welding.

DC transformers have a simple design with two main components: a transformer and a rectifier. The transformer produces and regulates AC entering the machine and the rectifier converts the AC to DC current. Another important component of the DC transformer-rectifier is the cooling fan. Since a lot of heat is generated during power conversion, the fan helps prevent the machine from overheating.

Unlike transformers, which regulate AC, rectifiers convert AC to DC. Some welding machines offer both AC and DC power. When a switch is flipped, the alternating current is directed through the rectifier and direct current is produced for welding. See Figure 1.

Inverter

The **inverter** converts high-voltage, low-amperage AC current by changing the frequency to low-voltage, high-amperage DC. It does this by running the current through a series of rectifiers, gating transistors, and transformers. The inverter is a popular power source for a number of reasons. Figure 2 shows an inverter type welder.

- It is lightweight and highly portable.
- It is economical to run because it uses much less electricity to generate the necessary power.
- It has fewer maintenance issues and fewer breakdowns.
- It produces a stable current, which is better for welding.
- It has versatile power input 110, 220, 480, single- or 3-phase.

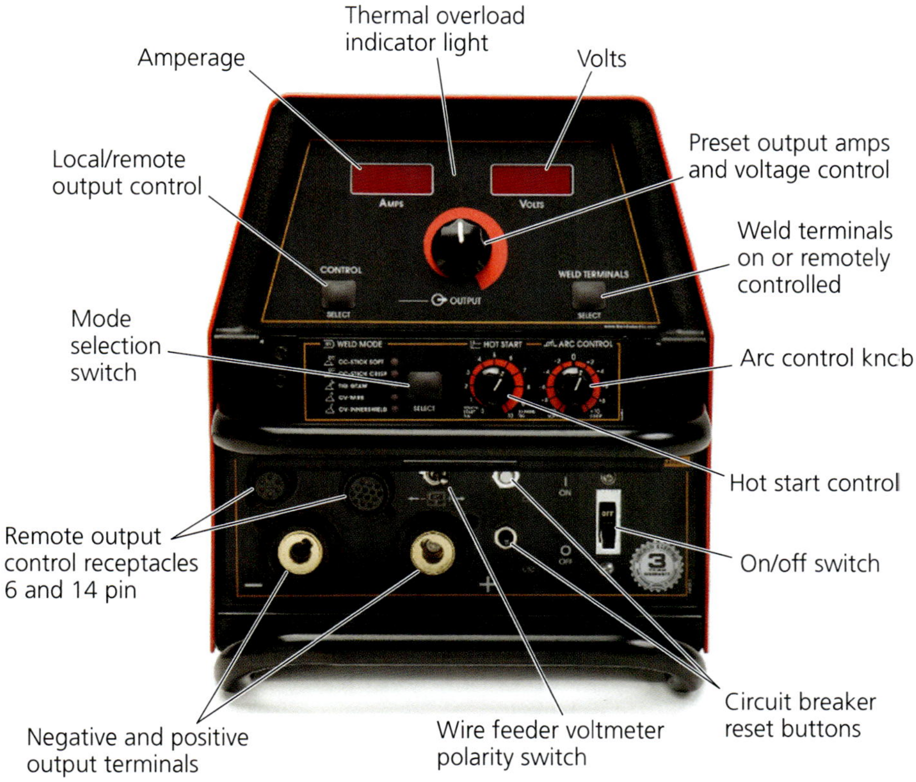

Figure 2
Inverter type welder

Generators

There are two types of motor generators; they are classified according to the motor that provides the power.

1. The first type of generator has an AC motor, a DC generator, and an exciter built on a single shaft. This type of motor is powered by AC electricity found on the jobsite.
2. The second type of generator, the alternator-motor-driven generator, generates its own electricity by running a gasoline- or diesel-powered alternator motor. See Figure 3. These welders are used when power is unavailable at a jobsite.

Figure 3 shows an engine-driven generator for stick welding.

Figure 3
Engine-driven generator for stick welding

Classifying Power Sources

The power sources discussed thus far have been identified by the pieces of equipment involved, but a power supply can also be classified as to whether it supplies *AC* or *DC* power, and it can be classified by its *output slope.*

Output Slope The **output slope** is the change in relationship between the output voltage and the output current (or amperage) of the machine as either the current or the welding workload is changed. This change is controlled by whether the power is constant current or constant voltage; it cannot be both.

Constant Current Current is measured in amps. Amps measure the amount of current flow exiting a machine, which determines how hot the weld can become. When the number of amps exiting the machine does not change, it is considered **constant current**. It is also known as variable voltage, because while the current remains constant, the voltage will vary as the welding load changes. GMAW and GTAW welders are the primary constant current welding machines. Constant current machines are available in both DC and AC welding current.

Constant Voltage When a welding machine produces **constant voltage**, the current will change depending upon the welding workload. It can be understood as the amount of pressure being exerted by an electric current.

Duty Cycle Another important concept is **duty cycle**, which is the amount of time a piece of equipment can operate at a certain output. If a welder had a 100 percent duty cycle, the machine could run at its rated current continuously for the duration of the welding job. However, most machines cannot run at 100 percent duty cycle. Duty cycles are measured in 10-minute increments. A welder with a 60 percent duty cycle can run continuously six minutes of every ten. After the six minutes are up, the machine must be stopped in order to cool. Some machines have only a 20 percent duty cycle and can only run two out of every ten minutes.

productivity tip

Select a welding machine with a duty cycle that will not require the welder to stop welding to let the machine cool.

Self Check

1. What are the two main components of a DC transformer?
2. Which two welding machines use constant current?
3. What is a duty cycle?

3 Electrode Holders

The **electrode holder** is a device that holds the electrode used for welding. This device is sometimes referred to as a stinger. There are two types of electrode holders: insulated and noninsulated. Insulation prevents accidental short circuits if the holder comes into contact with the work.

Electrode Holder Components

The traditional electrode holder design is comprised of nine parts. See Figure 4.

One drawback of traditional electrode holders is that the electrode has to be rather long in order to be useful. New designs allow a much shorter electrode to be used. Figure 5 shows a new electrode holder design with a twist-lock device. This holder has a strong clamping force, thereby allowing the electrode to remain secure, even when used to break through slag to start an arc. Figure 6 shows an electrode holder with an angled head. This type of holder is also lighter and shorter than other holders of the same capacity.

Electrode holders come in different sizes and are rated for different amperages. Welding equipment manufacturers will recommend a specific rated electrode holder based on the amperage of the welding machine and the welding job requirements. If the holder used is too small for the machine, overheating can occur.

Figure 4
Components of an electrode holder

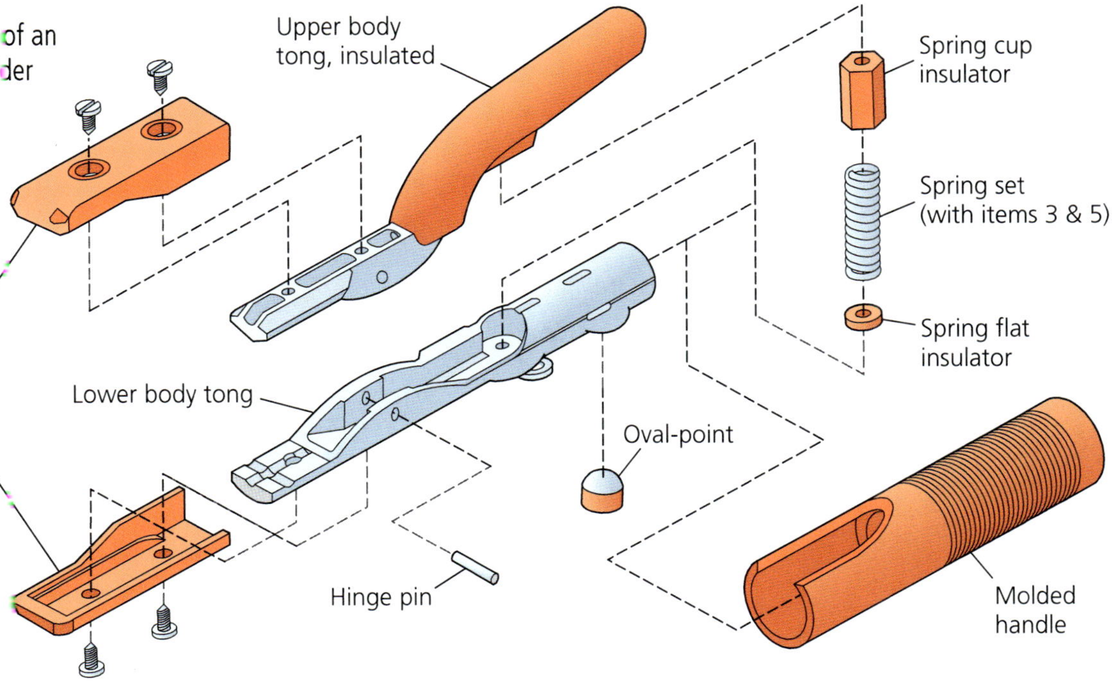

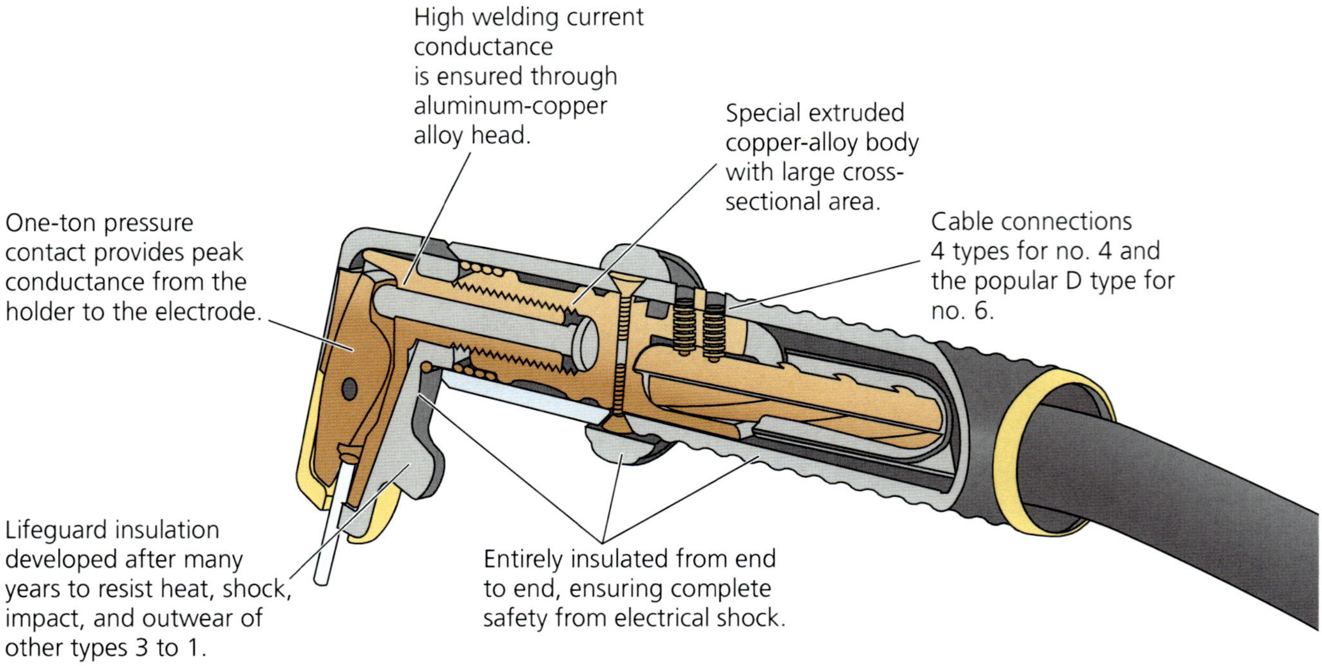

Figure 5
Electrode holder with a twist lock

Figure 6
Electrode holder with angled head

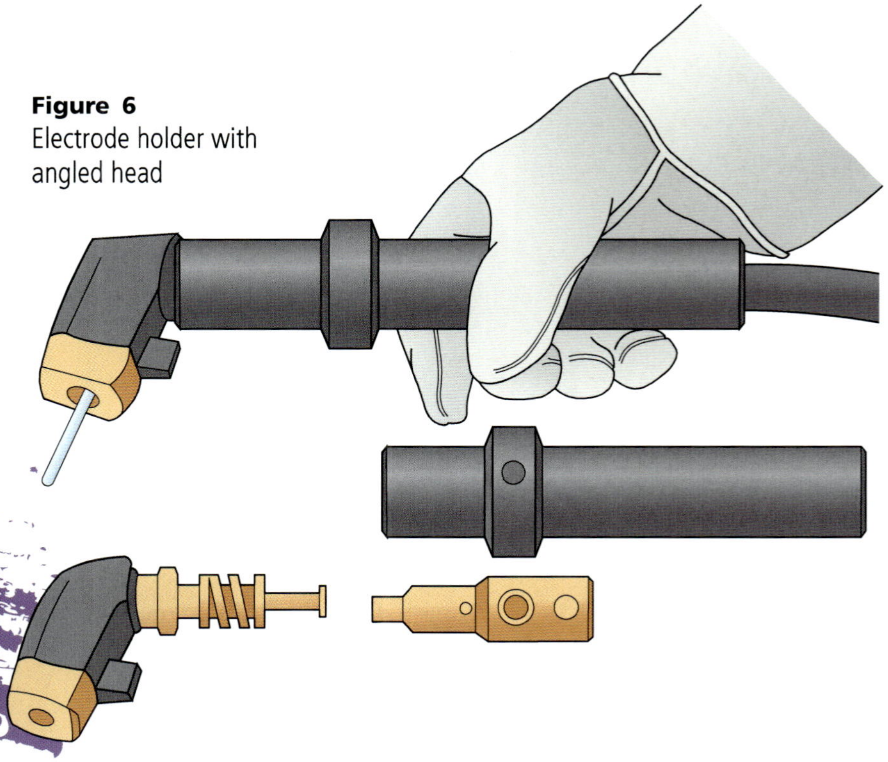

It is important to make certain each component of the electrode holder is functioning properly.

Inspection

It is important to make certain each component of the electrode holder is functioning properly. For instance, if the lower body tong has a damaged hinge, it might not hold the electrode firmly. Also, if any insulation is missing from the handle, the welder could be burned.

4 Work Clamps

It is necessary to have a good **work clamp** to make a good weld. Work clamps are used to ground the work to the welding machine. If the ground is loose, the welder may not produce enough heat to make a good weld and the machine itself could be damaged.

Grounding the Welding Machine

There are many styles of work clamps that ground the work to the machine.

- Use a work clamp or a spring-loaded clamp. See Figure 7 and Figure 8.
- Bolt or tack-weld the end of the ground cable to the work. See Figure 9.

Inspecting the Electrode Handle and Work Clamp

Before welding begins, the electrode holder handle must be inspected for wear, tear, and defects. The electrode must fit firmly within the grips and the grip must be fully insulated so that nothing is exposed that can conduct heat to the welder's hand.

The cable should be held firmly in whichever clamp is being used. Make sure that the clamp cannot be twisted or easily knocked out of place. Bolting or tack-welding the end of the ground cable to the workpiece ensures a permanent common ground.

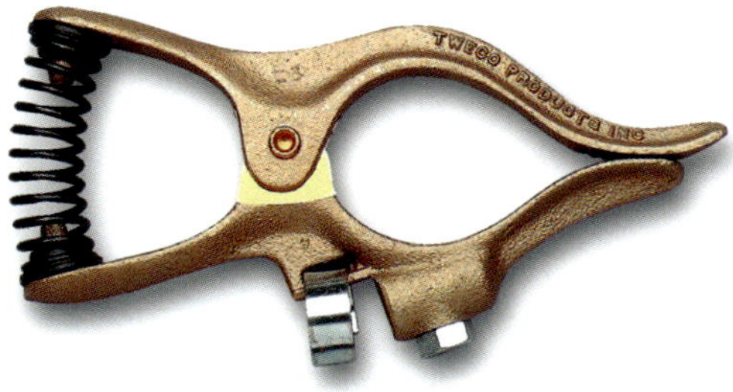

Figure 8
Spring-loaded work clamp

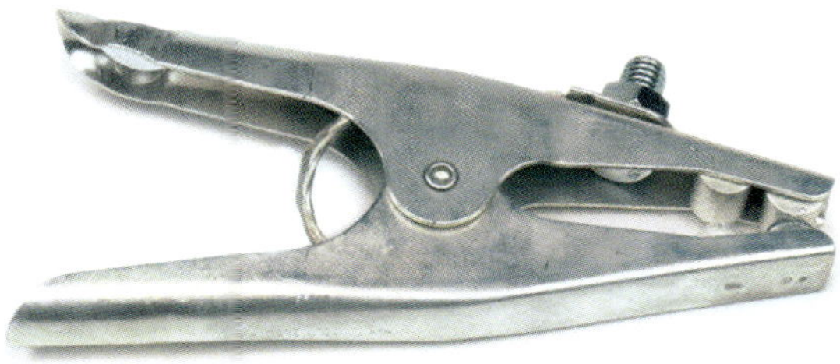

Figure 7
Work clamp

Figure 9
Bolted work clamp

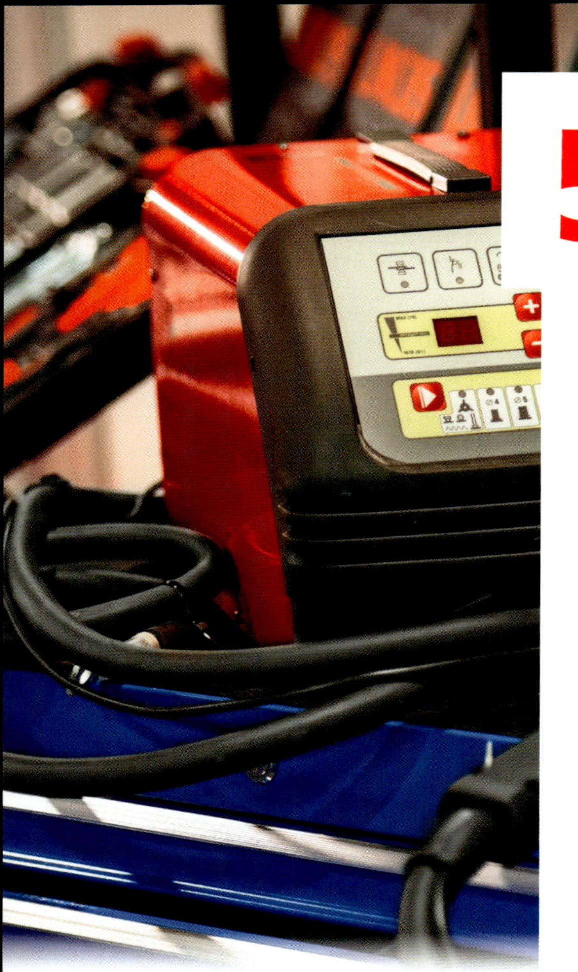

5 Cable Connections and Inspections

Cables are an important part of the welder's equipment. They must be properly connected and in good condition to ensure a good weld and a safe welder.

Cables must also be the right length and gauge for a good weld to be made. Cables that are too small for the welding machine's amperage could result in overheating and a poor weld. Cables that are too long can cause a voltage drop and negatively affect welding activity.

Cable Connections

Making solid cable connections is a critical first step to making a good weld. Solid connections are also necessary to ensure the welding machine's safe operation. There are several different types of cable connections. It is essential to understand the different types of connectors in order to match the machine to the cables. Figure 10 shows dense type connectors.

Cable Inspection

Since cables are used to complete the electric circuit that allows welding to occur, it is very important that they be inspected before work begins. There are three main things to check on cables.

Figure 10 Dense type connectors

1. Check to see if the cable to be used is in good shape; make sure there are no insulation tears or exposed copper wires.
2. Make sure the cable is the right size and length for the work being performed.
3. Check the lugs, work clamp, and electrode handle connections to ensure they are tight, there are no broken cable wires at the connection, and they make good contact with the copper wire in the cables.

Cables that are torn, frayed, or that have damaged lugs can affect the quality of the weld and also present an electrocution hazard. If the cable itself is in good shape, make sure the ground connection is secure and not likely to be knocked out of place. See Figure 11.

Self Check

1. What are the three methods of securing a ground?
2. What size electrode holder should be used?
3. Why are good connections important in the welding process?

Figure 11
Example of faulty cable

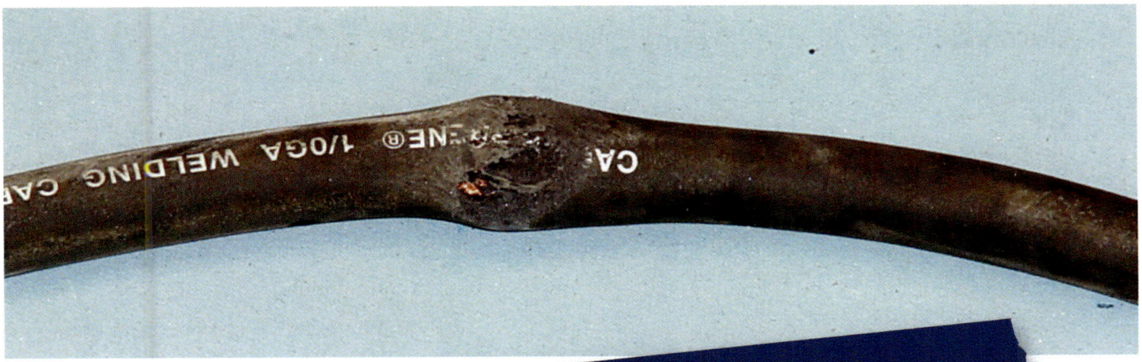

Making solid cable connections is a critical first step...

6 Hand Tools

Hand tools are a vital piece of a welding project. Projects must be laid out and the base metal must be cleaned before welding begins. Welds must be cleaned between passes and upon completion of the welding project.

Chipping Hammers

The **chipping hammer**, sometimes referred to as the cleaning hammer, is a required tool for welding. The chipping hammer is used to remove slag, rust, and other materials from the welded surface. This becomes critical when multiple-pass welds are made.

Chipping hammers are available as a hand tool or a power tool. Given the type and size of work being performed, one may be more suitable than the other.

Hand Chipping Hammers Most chipping hammers have two heads. One head is to chip rust and slag away from easily accessible areas of work. The other end is shaped like a pick, so that hard-to-reach areas can be cleaned. The chipping heads come in various sizes and shapes. They also come in combination with a wire brush so one tool serves two functions.

The distinguishing feature of most hand chippers is the handle. Some handles are made of wood and resemble a traditional hammer handle. See Figure 12. However, some hammers feature a steel spring handle. See Figure 13. These handles are designed to absorb the shock created when hitting a metal surface.

trade tip

To reduce hammer marks, do not use a chipping hammer until the metal has cooled to the point where it does not glow.

Figure 12
Traditional chipping hammer

Power Chipping Hammers Power chipping hammers come in electric, hydraulic, or pneumatic versions. Depending on the power situation on a jobsite and the type of welding being performed, one of these may be more suitable than the other.

Power chippers are good when there is a lot of cleaning and preparation to do. They are ruggedly built to withstand heavy use.

Wire Brushes

Like chipping hammers, **wire brushes** can be used to scrub away rust, slag, or other materials from welding work. They are used to clean and polish both before and after welding. See Figure 14.

Figure 13
Spring-handled chipping hammer

Wire brushes come in various shapes and sizes. Some wire brushes are even built into chipping hammers for efficiency. Wire brushes can also come with scrapers. The wires on the brush can be made out of brass, steel, or stainless steel.

Figure 14
Wire brush

Other Hand Tools: Soapstones and Grinders

Soapstone is a material made of talc, and it is used for marking metals. Soapstones come in different shapes to allow the welder to choose those best suited for a particular project. See Figure 15.

A grinder is a mechanical tool used to remove excess material or coatings, or to polish materials. Grinders come in a variety of sizes. A 5″ grinder is a popular size used by welders. Grinders can be used with many different wheels. See Figure 16.

Summary

There are three main types of welding machines: resistance welders, oxyfuel welders, and arc welders. Power sources for welding are AC-DC transformer, the DC transformer-rectifier, the inverter, and plug-in or alternator-motor-driven generators.

Additional required equipment for welding machines are the electrode holder, the work clamp, and cables. There are different styles and sizes available for each. Which one to choose depends on the type of work being done and the machine being used.

Always inspect welding equipment before beginning to weld. Before or after welding, a chipping hammer and wire brush are needed to clean slag, rust, and other materials from the weld.

Figure 15
Soapstone in a holder

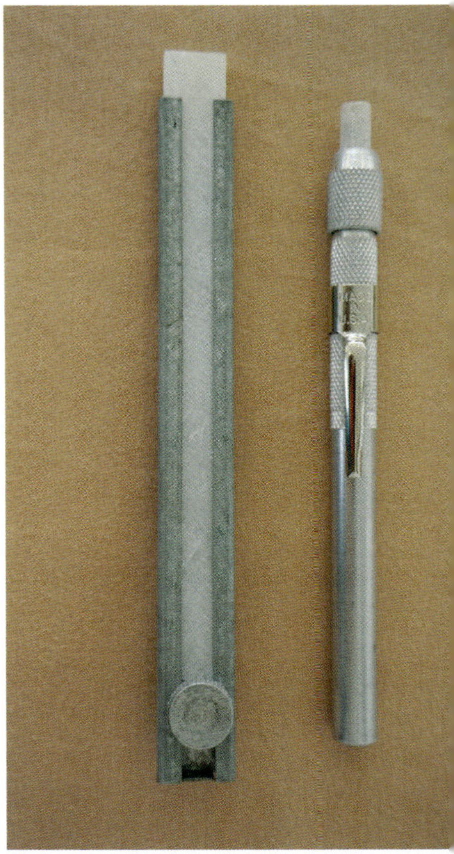

Figure 16
Grinder with grinding wheel attached and optional wire wheel for cleaning

Chapter Review

Show your understanding of the information in this chapter by answering the questions and filling in the blanks below.

1. Shielded metal arc welding is commonly known as ______________________ ______________________ or ______________________.

2. Inverters convert which of the following?
   a. high volt, low amp to low volt, high amp
   b. DC to amps
   c. AC to volts
   d. DCEN to DCEP

3. A DC rectifier takes ______________________ current and converts it to DC.

4. A MIG welder has a 70 percent duty cycle. This means that it can run:
   a. 8 minutes out of 10 minutes
   b. 2 minutes out of 10 minutes
   c. 7 seconds of every 10 minutes
   d. 7 minutes of every 10 minutes

5. Two types of electrode holders are ______________________ and ______________________.

6. Insulated electrode holders prevent accidental short circuits. (True; False)

7. Why is it important to ground the welding machine's AC power supply?

   ____

   ____

8. What are the different ways to ground the welding circuit?

   ____

   ____

9. Why should a welder inspect the equipment before use? Give some examples of what might be found.

   ____

   ____

   ____

10. Cables should be of sufficient size to handle the amount of ______________________ flowing through them.

11. Chipping hammers are used to clean______________________, ______________________, and other material from welding work.

12. Some chipping hammers come with a spring handle. This handle helps absorb:
    a. sweat
    b. power
    c. shock
    d. work

13. Power chipping hammers can be ______________________, ______________________, or ______________________.

14. Wire brushes are used to clean ______________________, ______________________, and other material from welded work.

15. Wire brushes can be made of:
    a. plastic
    b. wood
    c. copper
    d. stainless steel

CHAPTER 5

Shielded Metal Arc Welding (SMAW)

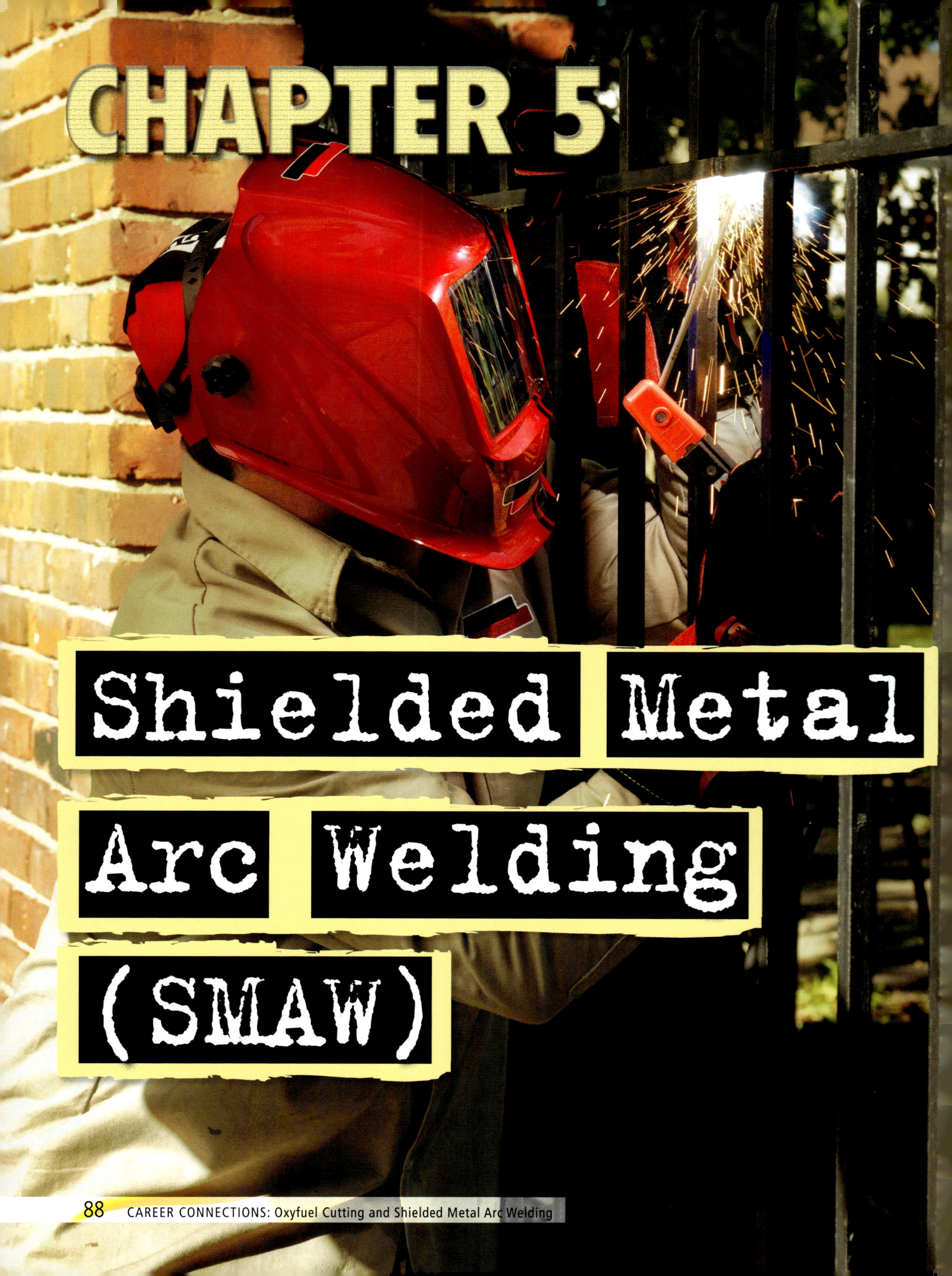

Shielded metal arc welding (SMAW), also known as stick welding, is perhaps the most widely used form of arc welding today. Shielded metal arc welding works on a wide range of metals. There are many important aspects of shielded metal arc welding to learn about, including safety issues, how the SMAW welding machine operates, and how to make good welds.

When welding, it is important to understand what the welding symbols mean to be able to effectively follow directions.

Contents

Objectives

- Describe the safety issues associated with shielded metal arc welding (SMAW).
- Describe how SMAW works.
- Describe electrodes used in SMAW.
- Identify the specific electrode classifications.
- Describe welding techniques used with this process.
- Demonstrate a knowledge of the types of electrodes to select.
- Describe how backing plates are used.
- Interpret weld bead examples.
- Demonstrate a knowledge of what the welding symbols mean.

Key Terms

alternating current (**AC**) electric current that reverses direction at regular intervals

backing plate used as support when creating a joint weld

bead or **weld bead** continuous deposit of fused metal

direct current (**DC**) electric current that flows in a constant direction

direct current electrode negative (**DCEN**) or **straight polarity** electrode lead is the negative side and the workpiece lead is the positive side of the circuit

direct current electrode positive (**DCEP**) or **reverse polarity** electrode lead is the positive side and the workpiece lead is the negative side of the circuit

electrode metal rod that is coated to provide an oxygen-free atmosphere, arc control, and a slag layer to protect the weld when cooling

inverter changes frequency to convert high-voltage, low-amperage AC current to low-voltage, high-amperage DC current

polarity direction in which current travels

shielded metal arc welding (**SMAW**) or **stick welding** welding process that uses an electric current (AC or DC) to form an arc between an electrode coated in flux and the metals to be joined

stinger common term for the electrode holder

stringer bead bead made with very small movements; should be no more than twice the width of the diameter of the electrode

weave bead bead made with small movements; should be no more than four times the width of the diameter of the electrode

1 Shielded Metal Arc Welding Safety

Shielded metal arc welding (SMAW) can be dangerous. Welders must follow safety guidelines below as well as the safety guidelines discussed in depth in the Welding and Cutting Safety chapter.

Personal Protective Equipment

To prevent contact with sparks and flying particles, welders should wear the following personal protective equipment (PPE):

- dry, insulated gloves that have no tears or holes
- protective garments such as leather gloves, a heavy shirt, cuff-less pants, shoes with high uppers or pull-on boots or boots with leather laces, and a cap
- approved safety goggles with side shields, in addition to wearing a welding hood or helmet with the minimum shade rating recommended for the procedure
- approved helmets or hoods that protect the face, neck, and ears
- leggings, aprons, sleeves, shoulder covers, and bibs
- flame-resistant earplugs, protective earmuffs, or both

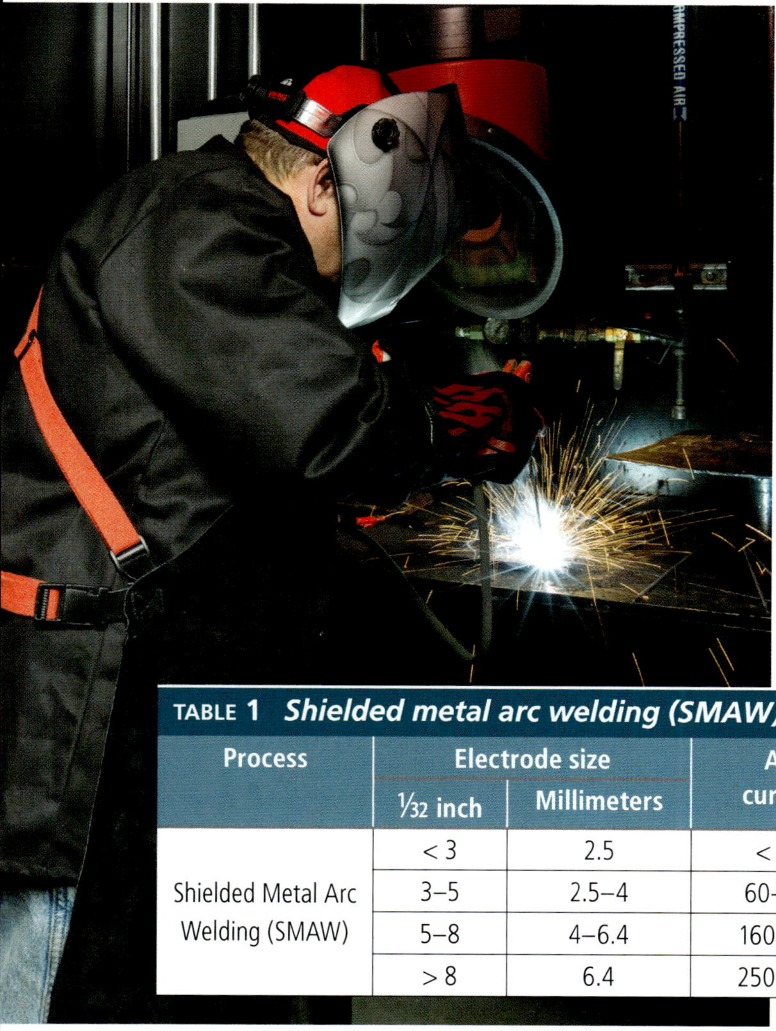

Shaded welding helmets should be worn while performing shielded metal arc welding. Table 1 lists the appropriate shading to wear for shielded metal arc welding as well as other welding processes. The welder should wear the recommended minimum shade during all welding activities. Anyone observing or working near the welding activity should also wear at least the minimum shade.

TABLE 1 ***Shielded metal arc welding (SMAW) guide for shade numbers***

Process	Electrode size		Arc current	Minimum protective shade	Suggested shade number (comfort)
	1/32 inch	Millimeters			
Shielded Metal Arc Welding (SMAW)	< 3	2.5	< 60	7	—
	3–5	2.5–4	60–160	8	10
	5–8	4–6.4	160–250	10	12
	> 8	6.4	250–550	11	14

2 What Is Shielded Metal Arc Welding (SMAW)?

In **shielded metal arc welding** (**SMAW**), commonly known as **stick welding** or *arc welding*, the weld occurs when the heat from the electric arc brings the metal to be welded and the electrode to a molten state. The heat melts the electrode and the work surface near it. The molten droplets from the electrode gather in a molten pool on the work area to create the weld. This molten pool is often called the puddle. The arc can reach temperatures above 9,000° F.

Shielded metal arc welding is used on *ferrous* and some *nonferrous* metals. It is especially effective on the following metals.

- carbon steel
- low-alloy steel
- stainless steel
- cast iron
- nickels and nickel alloys
- aluminum and aluminum alloys

Figure 1 shows a generator-welder set up for SMAW.

A visual explanation of the heating, melting, and welding area is shown in Figure 2.

Figure 1
Gas-powered SMAW welding machine

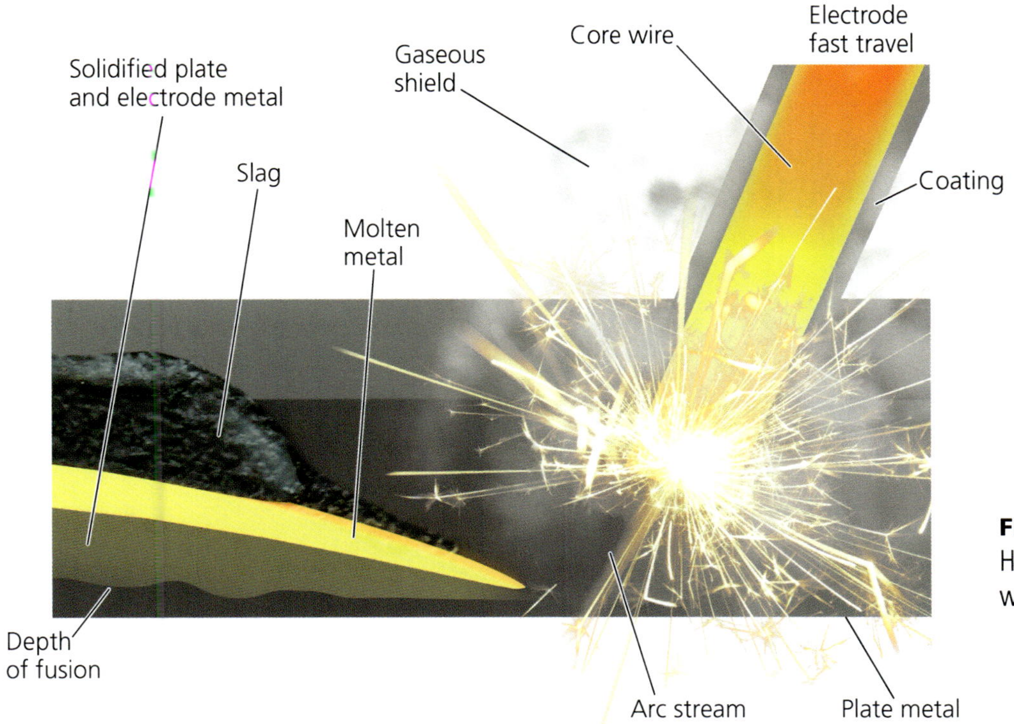

Figure 2
How SMAW works

Shielded metal arc welding is extremely portable. It can be performed indoors, outdoors, and in any position. It is used in a wide variety of trades and industrial settings and maintenance operations.

Welding Essentials

There are five welding essentials. Each of these must be present to generate a good quality weld.

- correct electrode
- correct amperage
- correct welding arc length
- correct travel speed
- correct electrode angle

In shielded metal arc welding, **electrodes** heat the workpiece and provide filler metal for the weld. It is important that the correct electrode for the work be used. If the incorrect electrode is used, the weld may not take or quickly fail due to incompatible metals. The position in which the welding is to be done affects the selection of electrodes. Certain electrodes work better in the flat position. Other electrodes work well in all positions.

The correct amperage is critical. If the amperage is incorrect, proper fusion may not be maintained and the metal may not melt.

The arc length is important as well, because if the electrode is held too close or too far from the work the weld will not be of good quality.

Travel speed is also important. Travel too quickly and the weld may be uneven. Travel too slowly and the weld will be too large.

The electrode must be held at the correct angle to weld most efficiently. In shielded metal arc welding, the electrode holder is commonly called a **stinger**.

Shielded metal arc welding is a manual process. The quality of the weld is in the hands of the welder. Figure 3 illustrates the versatility of the welding technique.

Figure 3
Welding with SMAW

Self Check

1. Where can shielded metal arc welding be used?
2. Why is it important to choose the correct electrode?

Power Sources

Welding power sources are also called power supplies and welding machines. There are different types of power sources used for shielded metal arc welding.

Transformers If AC power is supplied at a jobsite, the voltage present will be too high for the machine. A *transformer* operates to reduce the AC voltage to a level that can be used by the AC welding unit. It is called a step-down transformer.

Rectifiers A *rectifier* is used when the power source is AC but the welding unit requires DC. The rectifier converts the current from AC to DC so that the welding machine can operate.

Inverters An **inverter** changes frequency to convert high-voltage, low-amperage AC current to low-voltage, high-amperage DC current. It does this by running the current through a series of rectifiers, gating transistors, and transformers.

Generators A generator produces DC power. Gas- or diesel-powered alternators or generators can be used where there is no other power source available.

There are five essentials of welding.

Polarity

In **direct current** (**DC**) electricity flows in a constant direction. The direction of the current is its **polarity**. The choice of polarity can affect the quality of the weld. Polarity should be chosen based on the type of metal and electrode being used.

Direct Current Electrode Positive (Reverse Polarity) In **direct current electrode positive** (**DCEP**), also called **reverse polarity**, the electrode lead is the positive side of the circuit. The ground lead is the negative. This is the most common form of shielded metal arc welding. It is called reverse polarity because the electrical current flows from the base metal to the electrode and the filler metal is deposited in the reverse direction—from the electrode to the base metal.

DCEP produces a deep, penetrating weld. Figure 4 shows a machine in DCEP mode.

Figure 4
Machine in DCEP

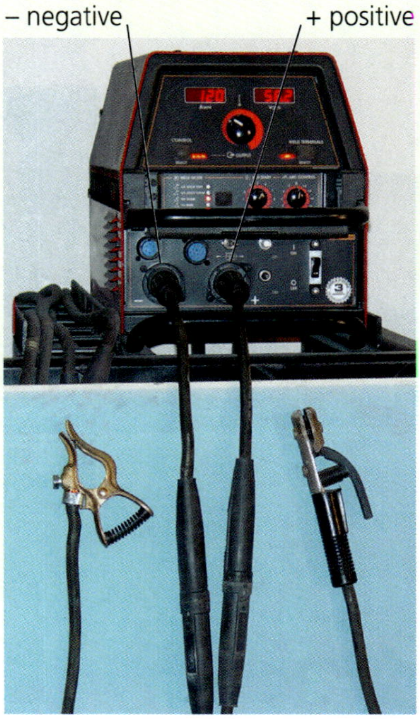

Direct Current Electrode Negative (Straight Polarity) In **direct current electrode negative** (**DCEN**), also called **straight polarity**, the electrode lead is the negative side and the work lead is the positive side of the circuit.

Figure 5
Workers setting a bridge structure

DCEN produces a shallow weld but allows for a greater rate of deposition than DCEP. In DCEN mode, the electrical current and the deposited filler metal flow in the same direction—from the electrode to the base metal. Figure 5 shows workers setting a bridge structure.

Alternating Current Changing the polarity on an AC welder does not have an effect on the weld because **alternating current (AC)** alternates from positive to negative. In SMAW, AC is sometimes used for welding aluminum. The most efficient current for shielded metal arc welding is DC, either DCEP or DCEN.

Polarity Switch Some shielded metal arc welding power sources have a polarity switch. This switch allows the welder to switch polarities, based on which one is best suited for the job. Welding machines without a polarity switch must be physically switched by switching the ends of the leads from the positive terminal to the negative terminal.

Self Check

1. What is the most common polarity used for SMAW?
2. How are polarities switched on a SMAW welding machine?

3 Electrode Identification

All electrodes use the same system of identification/classification developed by the American Welding Society (AWS). Covered arc welding electrodes have the classification number printed or stamped on the electrode covering within 2 ½″ of the base of the electrode.

Each number and letter in the classification system has a meaning. The following figures and tables provide the breakdown of the coding system.

Figures 6 and 7 illustrate the electrode numbering system.

See Table 2 for a listing of the welding characteristics of mild steel electrodes.

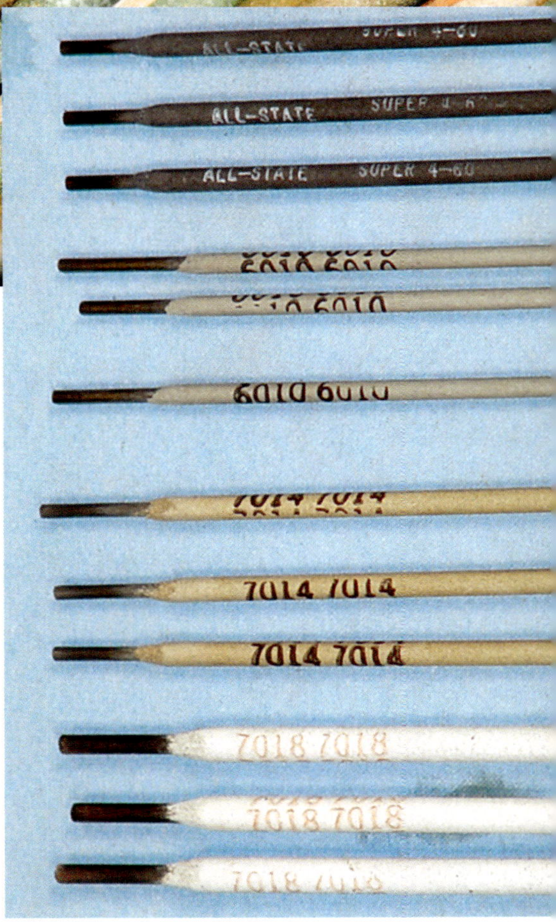

Figure 6
Location of the electrode classification number

Figure 7
AWS SMAW rod classifications

AWS A5.1 Carbon Steel Electrode

E6010

- Electrode (E)
- Tensile in ksi (60)
- Position (1)
- Type of coating and current (0)

Position

1 – Flat, horizontal, vertical, overhead
2 – Flat and horizontal only
4 – Flat, horizontal, vertical down, overhead

Types of Coating and Current

Digit	Type of Coating	Current
0	Cellulose sodium	DC+
1	Cellulose potassium	AC, DC±
2	Titania sodium	AC, DC–
3	Titania potassium	AC, DC +
4	Iron powder titania	AC, DC±
5	Low hydrogen sodium	DC+
6	Low hydrogen potassium	AC, DC+
7	Iron powder iron oxide	AC, DC±
8	Iron powder low hydrogen	AC, DC±

AWS A5.5 Low Alloy Steel Electrode

E8018-B1 H4R

- Electrode (E)
- Tensile in ksi (80)
- Position (1)
- For AC or DC+ (8)
- Chemical composition (B1)
- Maximum diffusible hydrogen level obtained on the AWS A4.3 test. (Example: H4 = max 4.0 ml/100 g) (H4)
- Moisture resistant. Meets specific low moisture pickup limits under controlled humidification tests. (R)

Chemical Composition of Weld Deposit

Suffix	%Mn	%Ni	%Cr	%Mo	%V
A1				.50	
B1			.50	.50	
B2			1.25	.50	
B3			2.25	1.00	
C1		2.50			
C2		3.25			
C3		1.00	.15	.35	
D1/D2	1.25-2.00			.25-.45	
G(1)		.50 min.	.30 min.	.20 min.	.10 min.

(1) Only one of the listed elements is required

TABLE 2 ***Welding characteristics of mild steel electrodes***

Electrode	Type of coating	Position of welding	Type of current	Penetration	Rate of deposition
E6010	High cellulose sodium	All positions	DCEP	Deep	Average
E6011	High cellulose potassium	All positions	DCEP, AC	Deep	Average
E6012	High titania sodium	All positions	DCEN, AC	Medium	Good
E6013	High titania potassium	All positions	DCEP, DCEN, AC	Mild	Good
E7014	Iron powder titania	All positions	DCEP, DCEN, AC	Medium	High
E7015	Low hydrogen sodium	All positions	DCEP	Mild to medium	Good
E7016	Low hydrogen potassium	All positions	DCEP, AC	Mild to medium	Good
E6020	High iron oxide	Flat horizontal fillets	Flat: DC, AC horizontal fillets: DCEN, AC	Deep	High
E7024	Iron powder titania	Flat horizontal fillets	DCEN, DCEP, AC	Mild	Very high
E6027	Iron powder iron oxide	Flat horizontal fillets	Flat: DC, AC horizontal fillets: DCEN, AC	Medium	Very high
E7018	Iron powder low hydrogen	All positions	DCEP, AC	Mild	High
E7028	Iron powder low hydrogen	Flat horizontal fillets	DCEP, AC	Mild	Very high
E7048	Iron powder low hydrogen	All positions vertical down	DCEP, AC	Mild	High

Appearance of bead	Spatter	Slag removal	Minimum tensile strength (psi)	Yield point (psi)	Minimum elongation in 2 inches (%)	Electrode
Rippled and flat	Moderate	Moderately easy	62,000	50,000	22	**E6010**
Rippled and flat	Moderate	Moderately easy	62,000	50,000	22	**E6011**
Smooth and convex	Slight	Easy	67,000	55,000	17	**E6012**
Smooth and flat to convex	Slight	Easy	67,000	55,000	17	**E6013**
Smooth and flat to convex	Slight	Easy	70,000	60,000	17	**E7014**
Smooth and convex	Slight	Moderately easy	70,000	60,000	22	**E7015**
Smooth and convex	Slight	Very easy	70,000	60,000	22	**E7016**
Smooth and flat to concave	Slight	Very easy	62,000	50,000	25	**E6020**
Smooth and slightly convex	Slight	Easy	72,000	60,000	17	**E7024**
Flat to concave	Slight	Easy	62,000	50,000	25	**E6027**
Smooth and flat to convex	Slight	Very easy	72,000	60,000	22	**E7018**
Smooth and slightly convex	Slight	Very easy	72,000	60,000	22	**E7028**
Smooth and slightly convex	Slight	Easy	72,000	58,000	22	**E7048**

Setting Welding Current

The diameter, length, and type of electrode have a direct bearing on which current should be used in shielded metal arc welding. Table 3 illustrates the correct current range based on the electrode indicated.

It is important to understand that the core of electrodes, no matter their classification, is the same basic core wire. The differentiating factor is the coating.

TABLE 3 ***Approximate electrode amperage settings****

		Fast Freeze E6010–E6011	Fill Freeze E6013–E7014	Fast Fill E7024–E7014	Low Hydrogen E7018
Diameter of Electrode		***Current Setting***	***Current Setting***	***Current Setting***	***Current Setting***
Inches	**Millimeters**	**Amperes**	**Amperes**	**Amperes**	**Amperes**
3/32 in	2.4mm	40–90	75–105	85–155	70–140
1/8 in	3.2mm	75–130	100–165	100–175	90–185
5/32 in	4.0mm	80–160	135–225	160–270	140–230
3/16 in	4.8mm	110–225	185–280	220–330	210–300
7/32 in	5.6mm	200–260	235–340	270–410	230–380
1/4 in	6.4mm	220–325	260–425	315–520	290–440

*Manufacturer's recommended setting should always be followed.

productivity tip

Electrode suppliers have booklets containing information about electrodes and welding machine settings. Every welder should carry an electrode booklet.

trade tip

A general rule for setting the correct amps based on the chosen electrode is: rod diameter converted to a decimal (but ignoring the decimal point), minus 20, equals approximate current setting. Example:

1/8″ (0.125″) rod = 125
− 20
105 amps

Specific Electrode Classifications

The following information decodes the classifications of common electrodes. These electrodes are for carbon steel, low hydrogen, and other covered electrodes.

Carbon Steel Electrodes Below is a list of common types of carbon steel electrodes.

- E6010: All-position, DCEP (fast-freeze type)
- E6011: All-position, alternating current and DCEP (fast-freeze type)
- E6012: All-position, alternating current and DCEN (fill-freeze type)
- E6013: All-position, alternating current and DCEN or DCEP (fill-freeze type)
- E7014: All-position, alternating current and DCEN or DCEP (fast-fill type)

Low-Hydrogen Electrodes Below is a list of common types of low-hydrogen electrodes.

- E7015: All-position, DCEP (low hydrogen)
- E7016: All-position, alternating current and DCEP (low hydrogen)
- E7018: All-position, alternating current and DCEP (low hydrogen, iron powder)

- E7028: Horizontal and flat position alternating current and DCEP (low hydrogen, iron powder).
- E7048: Flat, horizontal, overhead, and vertical down positions, alternating current and DCEP (low hydrogen, iron powder)

Other Covered Electrodes Below is a list of common types of other covered electrodes.

- E6020: Horizontal fillet and flat position, alternating current and DCEN or DCEP, high iron oxide
- E7024: Horizontal fillet and flat position, alternating current and DCEN or DCEP (iron powder, titanium)
- E6027: Horizontal fillet welds and flat position, alternating current and DCEN or DCEP (iron powder, iron oxide)

Self Check

1. How does the electrode classification system work?
2. What are some factors that determine which electrode is used?

4 Factors in Shielded Metal Arc Welding

Knowing the basic techniques of shielded metal arc welding is essential before beginning to practice in the classroom or the lab. Below are some of the factors that a welder must understand in order to make good welds.

Choosing the Correct Electrode The choice of electrode should be based on several factors: the metal to be welded, the position of the metal, and the thickness of the metal. Refer to Tables 2 and 3 to help guide the electrode choice.

Correct Amperage The current setting should be chosen based on the choice of electrode. Rod diameter converted to a decimal, minus 20, equals the approximate current setting. See Table 3 above to find the correct current setting.

Rod Angle One of the factors that determines the quality of the weld is the angle of the electrode. The angle at which an electrode is held affects the surface tension of the metal, as does the force of gravity. Holding the electrode at the correct angle allows the weld to be uniform and of a better quality. Typically, the rod will be held at between a 10° angle and a 20° angle in the direction of travel when depositing the weld.

Arc Length Shielded metal arc welding requires a short, consistent arc length. The shorter distance from the work allows the heat to concentrate more efficiently and it reduces the effect of *arc blow,* a wandering of the arc stream due to fluctuation of the magnetic field. The closer distance also protects the weld from being damaged or destroyed by the surrounding air.

The arc distance depends on the electrode type and size, the welding position, and the amount of current. A welder can tell if the arc is the correct length by how well the metal is being transferred. A short arc makes a sharp crackling sound, while a long arc hisses and explodes at steady intervals.

Travel Speed Another factor that determines the quality of the weld is travel speed. Travel speed in turn depends on a number of elements, including the following.

- type of welding current, amps, and polarity
- position of welding (flat, vertical, overhead)

- melting rate of electrode
- thickness of metal
- surface condition of the base metal
- electrode manipulation

A sixth factor for a good weld is *slag*. Although one might think it is only an irritating by-product that must be removed, it actually performs a vital function during the welding process.

Slag A combination of the heated steel, the electrode, and the electrode covering is referred to as slag. It protects the electrode deposit (weld) from the air during cooling; it must be chipped or brushed off once the weld has cooled.

Self Check

1. What is the importance of slag?
2. What is the typical angle at which an electrode should be held while performing SMAW?

Striking an Arc

To begin the welding process, the welder must first strike an arc. Once the ground connection is connected to the work, lightly tap the workpiece with the tip of the electrode to perform a *tap start*. An arc can also be struck by scratching the workpiece with the tip of the electrode as well—the *scratch start*. This movement is similar to striking a match.

Figure 8 shows the tapping and scratching methods of striking an arc. The electrode will have a tendency to stick to the work upon touching it. If the electrode does stick, quickly break it free by wiggling it loose, not by pulling.

trade tip

The arc length should be a maximum of one rod diameter when welding flat position. When welding out of position, shorten the arc length slightly.

Weld beads and Electrode Manipulation

The **bead**, or **weld bead**, is a continuous deposit of fused metal. There are two types of bead: a stringer bead and a weave bead. These are created by the manipulation of the electrode.

Figure 8
Striking an arc

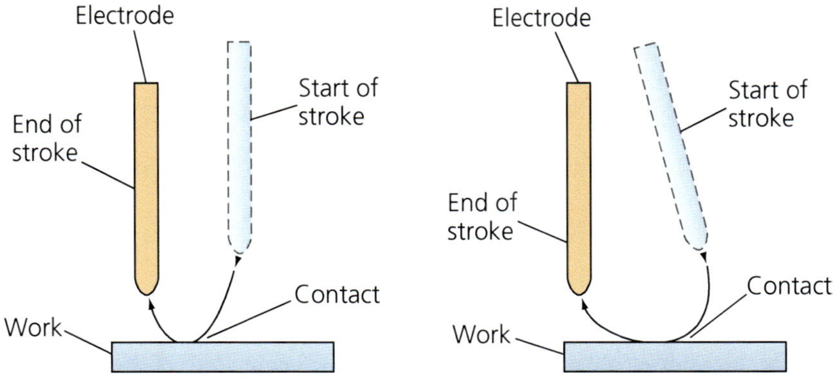

Striking the arc with a vertical down and up stroke.

Striking the arc with a scratching motion.

A **stringer bead** is made with very small movements, and should be no more than twice the width of the diameter of the electrode. A **weave bead** is produced by slightly larger movements of the electrode, and should be no more than four times the width of the diameter of the electrode.

Electrodes can be manipulated in various ways. Although these motions do not work well with fast fill or "drag" rods, the three basic motions when working with electrodes are the following.

- whip and pause
- crescent
- circular motion

Whip and Pause In the whip and pause method, the welder strikes an arc, builds a puddle, and whips the electrode out of the puddle in the direction of travel. The welder then whips it back into the leading edge of the first puddle, builds another puddle, and whips the electrode out of the puddle in the direction of travel. This process is repeated, maintaining uniform size and shape of the consecutive puddles, for the length of the weld. See Figure 9A.

The whip and pause technique should only produce stringer beads.

Crescent In the crescent method, the welder strikes an arc, builds a puddle, and moves the electrode out across the direction of travel and back to the puddle in a motion that moves the electrode about a quarter of the way along the perimeter of a small circle before returning to the puddle. This motion is repeated, all the while moving forward in the direction of travel approximately 1⁄16″ as each crescent is inscribed by the electrode. Do this repeatedly, keeping the size and shape of the crescent uniform. A single puddle will be maintained. See Figure 9B.

The crescent technique can produce either a stringer bead or a wash bead.

trade tip

The whip and pause, crescent, and circular welds are used as a cleaning action and for filling large gaps used with fast freeze rods.

Circular Motion Strike an arc, build a puddle, and move the electrode in the direction of travel about ¼″ using a circular motion. Repeat this motion while moving forward in the direction of travel, approximately 1⁄16″ of an inch with each circle. If the motion begins by moving the electrode clockwise, it must continue in a clockwise direction; if the motion begins counterclockwise, it must continue in a counterclockwise direction. Do this repeatedly, keeping the circular motion uniform. A single puddle will be maintained. See Figure 9C.

The circular motion technique will produce either a stringer bead or a weave bead, depending on the diameter of the circle.

Weld Positions

There are four basic positions in which welding can be performed: flat, horizontal, vertical up and vertical down, and overhead. Which electrode is used depends in a large part to which position the welding will occur.

Some electrodes are better suited to horizontal and flat while others are more suitable for overhead welding. Check the electrode codes to determine which one should be used.

The position in which the welding occurs also will impact the choice of current as well. It is more difficult to weld in the overhead position using AC current. This is due to the longer arc needed.

The two main types of weld joints ... are fillets and grooves.

Weld Progression

The direction of the weld is influenced by the weld position. For instance, if welding fast freeze rods in the vertical position, it is easier to start from top to bottom. For structural welding, however, vertical up is the method recognized. This is to keep out slag inclusions that occur using the vertical down technique.

When welding on a flat or horizontal surface, the welder should choose the direction that best suits the job at hand.

Welding Joints

The two main types of weld joint designs that need to be understood are fillets and grooves. Fillet joints are formed when two pieces of metal are placed approximately at a right angle to each other. Two common fillet

Figure 9
(A) Whip and pause
(B) Crescent
(C) Circular motion

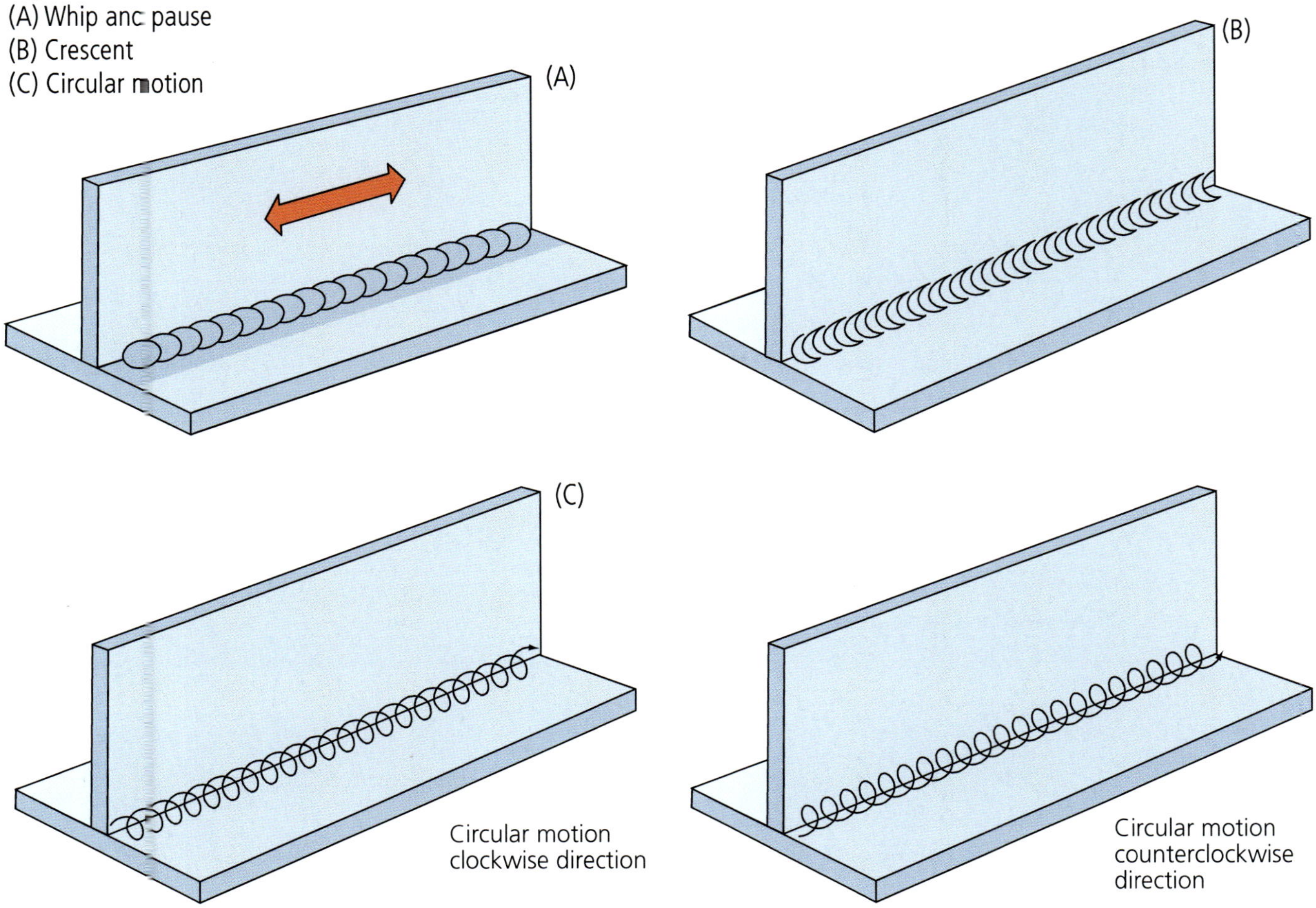

joints are lap and T-joints. A lap joint is formed when two pieces are laid on top of each other, forming a 90° angle at their edges. T-joints have the shape of a T. Groove joints are formed when two edges or surfaces are placed together with a gap in between them. Both grooves and fillets can be welded in any position. See Figure 10A and 10B.

When creating a joint weld in shielded metal arc welding, a **backing plate** is used. The backing plate is attached during the first weld pass. The backing plate should always be made of a metal compatible with the base metal. The backing plate must be free of slag and fit properly.

Self Check

1. What can happen to an electrode during arc striking?
2. Describe the three basic motions used when working with electrodes.

SMAW Weld Beads

Shielded metal arc weld beads vary according to the direction of the weld, the power usage, the chosen electrode, and the skill of the welder. Figure 11 shows what different beads look like under different conditions.

In Figure 11, the elevation and plan views of welds are shown under the following conditions.

- A shows current, voltage, and speed normal.
- B shows current too low.
- C shows current too high.
- D shows voltage too low, short arc.
- E shows voltage too high, long arc.
- F shows speed of travel too low.
- G shows speed of travel too high.

Figure 11
Understanding poor welds

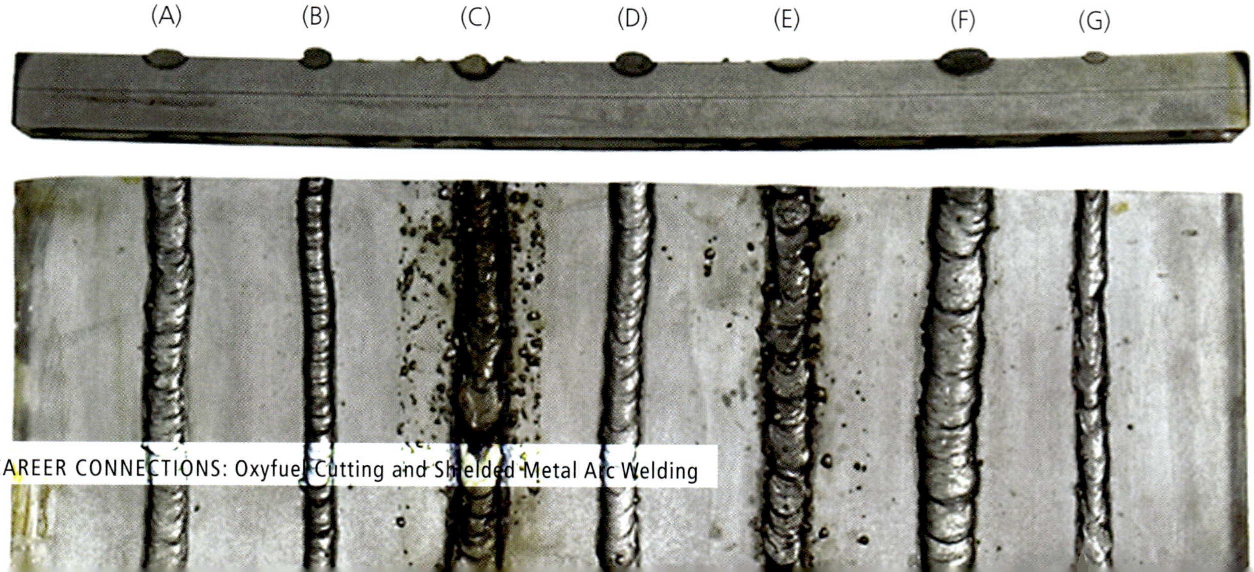

Figure 10
(A) Positions of test plate for fillet welds
(B) Positions of test plate for plate groove welds

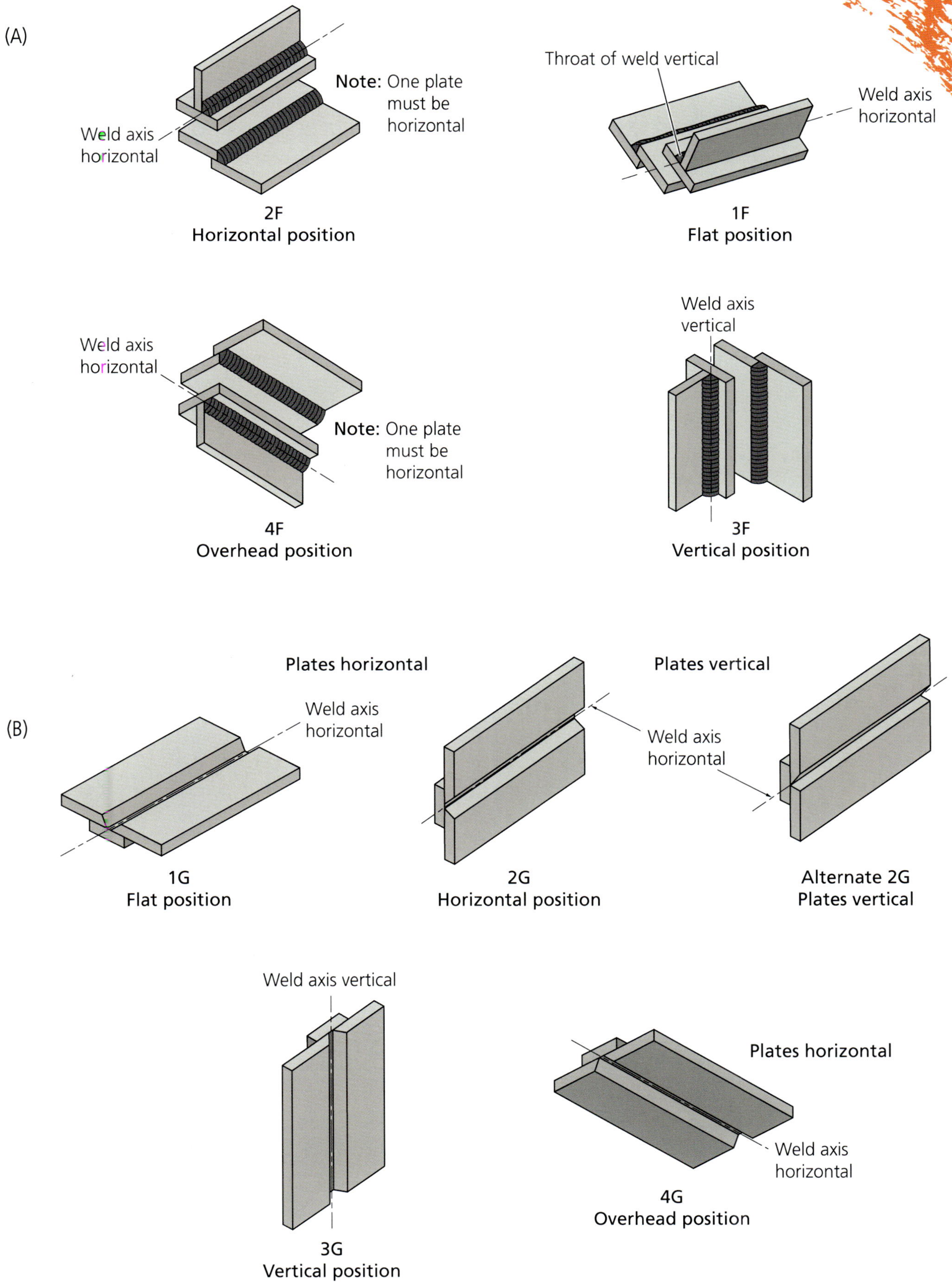

5 Weld Symbols

There are 16 basic weld symbols. These symbols are illustrated in Figure 12.

Figure 12
Weld symbols

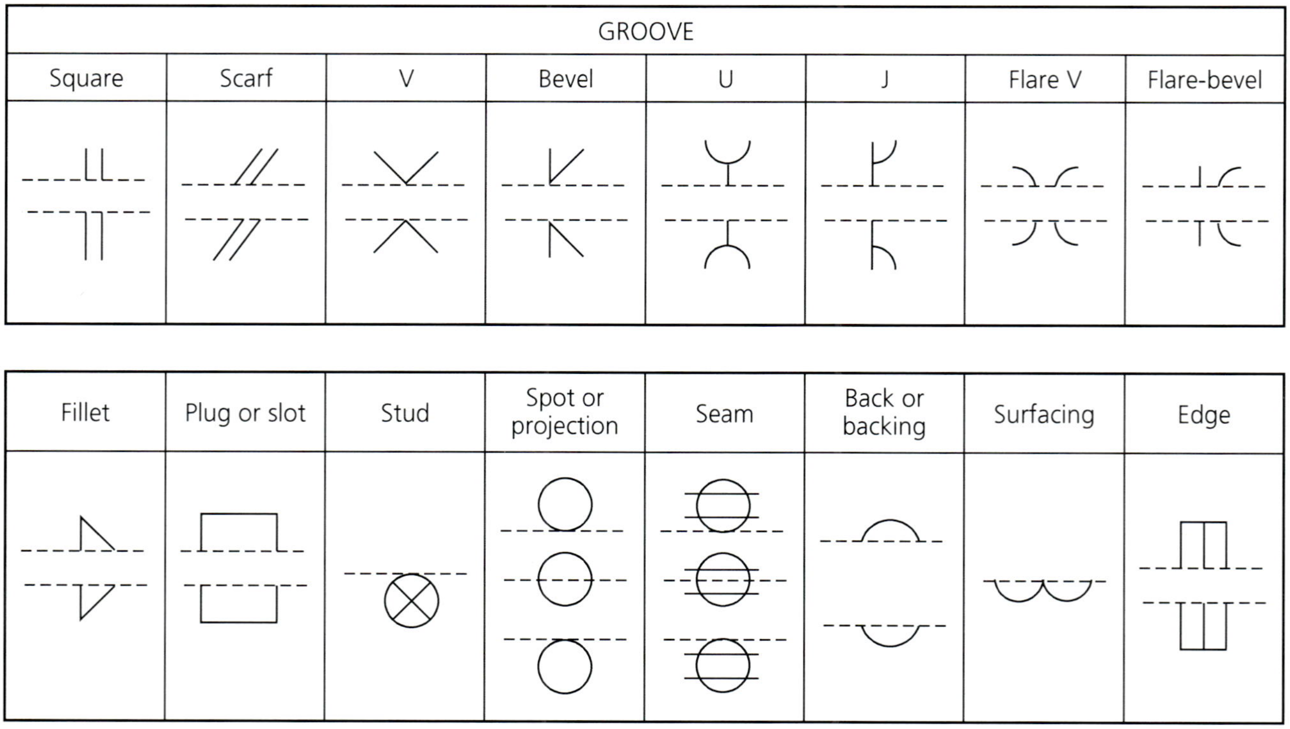

Summary

Shielded metal arc welding is probably the most widely used form of arc welding today. It is easy to master and works on a wide range of metals. Knowing how to operate the shielded metal arc welding machine correctly and safely will help increase welding quality.

Chapter Review

Show your understanding of the information in this chapter by answering the questions and filling in the blanks below.

1. The first two numbers in the electrodes classification stand for ____________________
____________________.

2. The choice of electrode should be based on the metal to be welded, the welding position, and the ____________________ of the metal.

3. DCEP stands for ____________________ ____________________
____________________ ____________________.

4. To change welding current polarity, the welding lead connections must be reversed at the positive and negative terminal posts on machines that don't have a polarity reversing switch. (True; False)

5. Which organization helped develop and maintains the electrode classification system?
____

6. The most efficient power source for SMAW is alternating current. (True; False)

7. ____________________ and ____________________ the electrode against the base metal are two ways to strike an arc.

8. The most difficult position to weld in is the ____________________ position.

9. Travel speed is not dependent on which of the following?
   a. position of welding (flat, upright, overhead, and so on)
   b. melting rate of electrode
   c. thickness of metal
   d. condition of ground cable

10. The four basic welding positions are ____________________, ____________________, ____________________, ____________________.

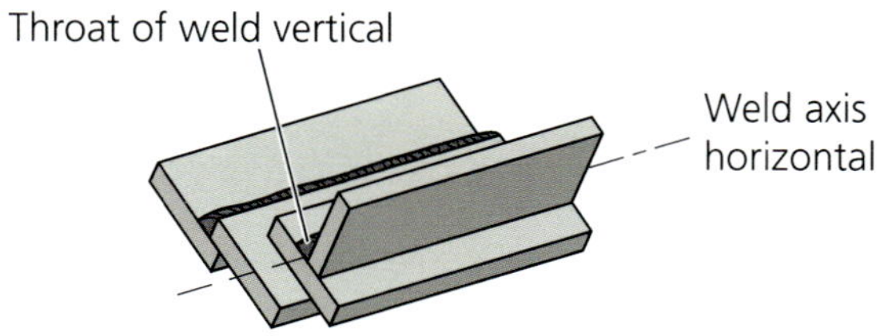

1F
Flat position

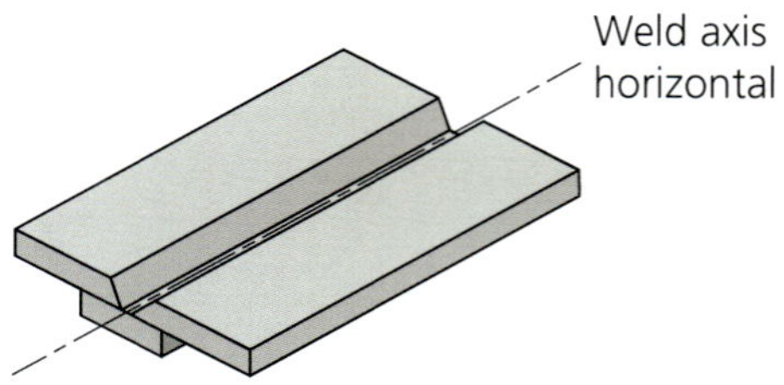

1G
Flat position

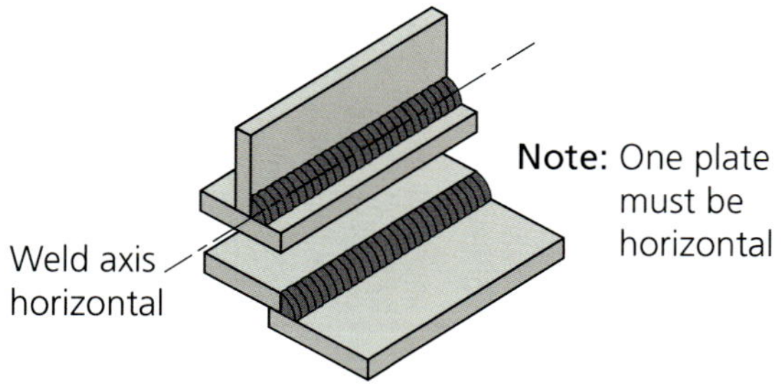

2F
Horizontal position

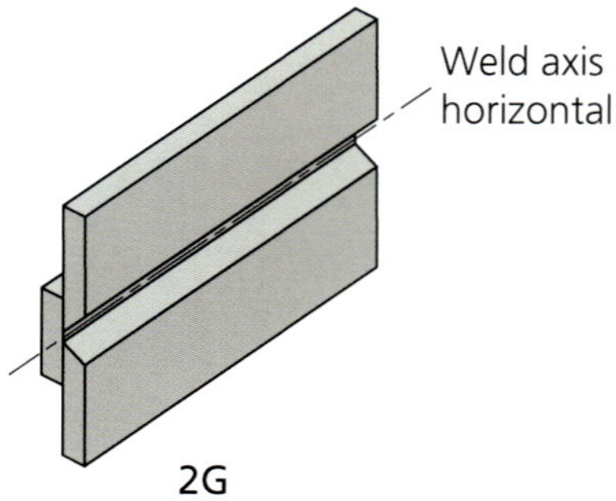

2G
Horizontal position

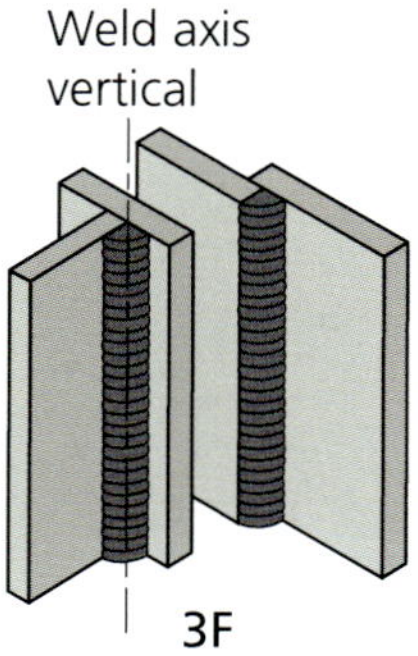

3F
Vertical position

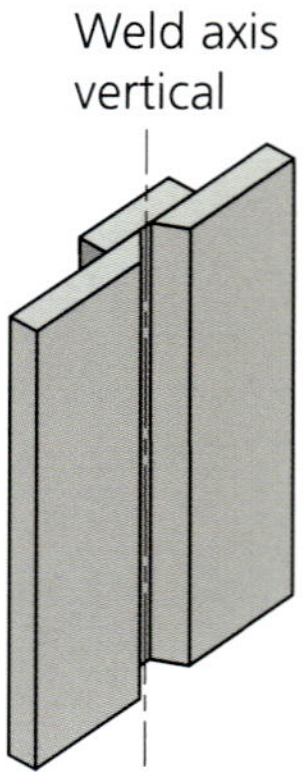

3G
Vertical position

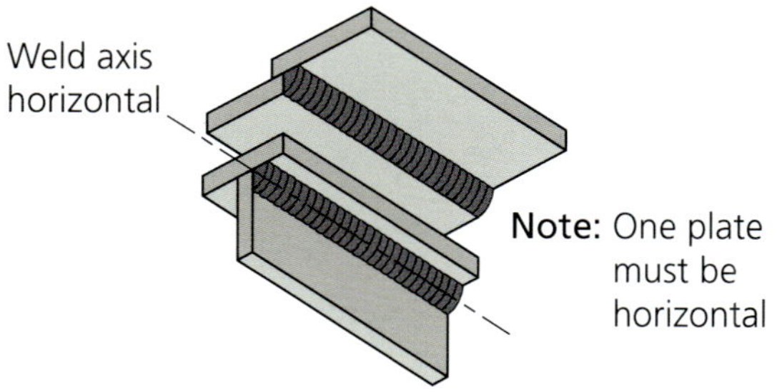

4F
Overhead position

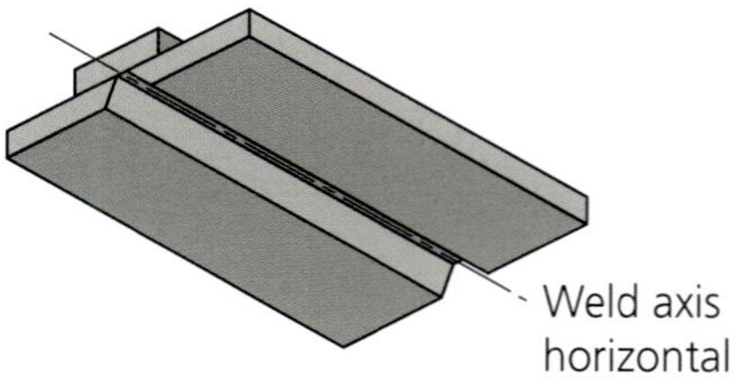

4G
Overhead position

Project 1

Fillet Welds With E6010 or E6011 Electrodes

Job Description

To complete this job successfully, the student should be able to assemble the project according to the given procedures and apply sequential weld beads with E6010 or E6011 electrodes, with minimal variation and defects. This will be done in the 1F, 2F, 3F, and 4F positions.

Materials
- two pieces of 1″ x 1″ x ¼″ angle iron cut 7″ long
- ⅛″ or 3/32″ E6010/E6011 electrode

Tools
- wire brush
- chipping hammer
- #11 vise-grips

Equipment
- SMAW welder
- grinder with a wire wheel and a grinding wheel

Safety Precautions

- Use the grinder in a safe manner, without injury to self or others.
- Ensure the equipment is in proper working order.
- Wear clear safety glasses at all times while in the shop.
- Wear appropriate clothing and work shoes at all times while in the shop.
- Wear the appropriate PPE while welding.
- Keep the work area clean and safe.

Procedure

1. Prepare and clean the surface to be welded.
2. Stand the two angles on their ends and lap the corners by ¼″, lining up the legs on the angles.
3. Clamp the angles together.
4. Attach the work clamp (ground) to the work surface or project.
5. Turn on the welder.
6. Set the welder to the proper amperage for the size and type of electrode being used. Start in the middle of the settings for the electrodes listed below.
   - E6010, 3/32″, 40–90 amperes
   - E6010, ⅛″, 75–130 amperes
   - E6011, 3/32″, 40–90 amperes
   - E6011, ⅛″, 75–130 amperes

Procedure, *cont.*

7. Insert the bare metal end of the electrode into the electrode holder (stinger).
8. Tack weld both ends of the angles with the electrode chosen.
9. Let the teacher check the fit-up.
10. Welding positions:
    - The flat (1F) position—the angles will be sitting on the table like an X.
    - The horizontal (2F) position—the angle irons will be orientated like a +. It will need to be put in a positioner or tacked to the table.
    - The vertical (3F) position—the angles will be standing on end with an open side facing the welder. It will need to be put in a positioner or tacked to the table.
    - The overhead (4F) position—the angle irons will be orientated like a +. It will need to be put in a positioner over the welders head.
11. In the flat and horizontal positions: if the welder is right-handed, weld left to right. If the welder is left-handed, weld right to left. This can also be used in the overhead position.
12. In the overhead position, the welder can also stand at the end of the angles and weld towards or away from his or her self.
13. In the vertical position, the welder is allowed to weld from the top to the bottom (Vertical Down), or from the bottom to the top (Vertical Up), of the angles. The bottom to the top is the preferred method.
14. Let the teacher check the position of the angles.
15. Use one of the electrode patterns listed in the chapter.
16. To avoid stopping and starting more than necessary, sit or stand in a position that allows the complete range of motion for the work piece.
17. Always remove the electrode from the electrode holder before hanging the holder up.
18. Never lay the electrode holder down on the table—with or without the electrode in it.
19. Using the scratch or tapping method, start the electrode and use the welders preferred technique to weld the root pass.
20. Weld the root bead, keeping it straight and uniform in appearance.
21. Remove any slag with a chipping hammer.
22. Clean the piece with a wire brush or grinder with a wire wheel.
23. The teacher will review the weld before continuing.
24. Remove any slag and clean each weld bead before welding the next weld bead in the sequence.
25. Continue to weld in the sequence shown in the figures on the next page; show the teacher every layer of weld beads for further instruction.
26. Weld stringer beads should be laid over half of each other; this is achieved by welding along the toe of the stringer bead before.
27. Dip the angle irons in a cooling tank every 2–3 weld layers.

28. When the one side of the angle is completed, let the teacher review the work.

29. Repeat the weld bead sequence on the opposite side if more practice is required by the teacher.

30. There should be equal amounts of weld on each leg of the angle.

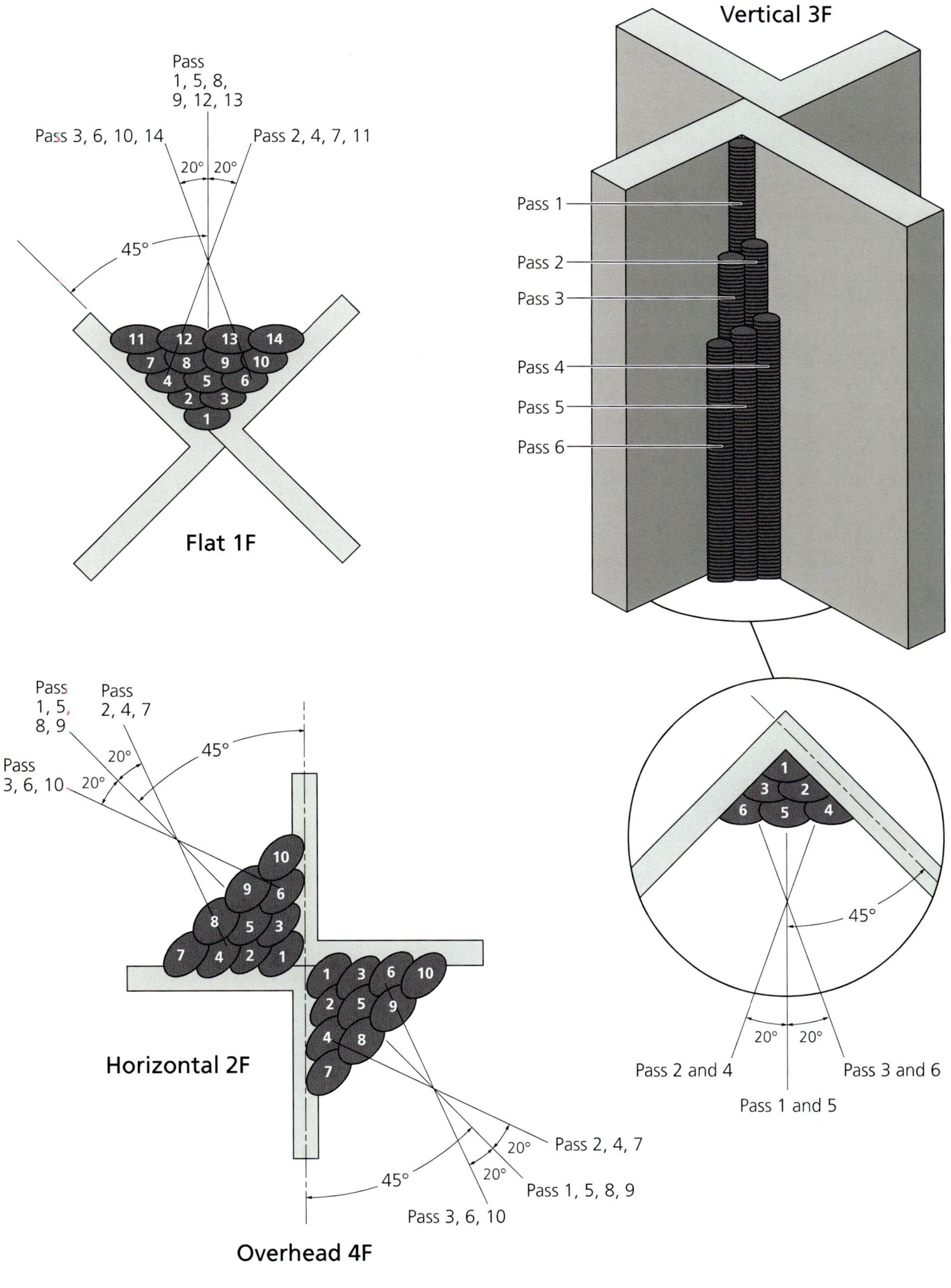

Fillet Welds With E6010 or E6011 Electrodes Rubric

POINTS:	5	4	3	2	1	0
Criteria:						
Pieces clean	Completely clean	¾ surfaces clean	½ surfaces clean	¼ surfaces clean	⅛ surfaces clean	Not Cleaned
Pieces tight	Tight	1/16″ gap	⅛″ gap			+⅛″
Pieces lap	¼″ lap	⅜″ lap	½″ lap	⅝″ lap		Assembled wrong
Proper polarity	Correct					Incorrect
Proper grounding	Lead tight to machine and work clamp tight to work piece	Lead tight to machine and work clamp is on coated surface of work piece	Lead tight to machine or work clamp is tight to work piece, not both	Work clamp not on work piece, lead tight to machine		Not grounded
Proper amperage	± 10 amps	± 15 amps	± 20 amps	± 25 amps	± 30 amps	± 35 amps
Appearance of weld bead face	0″–1″ Bad appearance	1″–3″ Bad appearance	3″–5″ Bad appearance	5″–7″ Bad appearance	7″– 9″ Bad appearance	+9″ Bad appearance
Undercut	No undercut	1″–2″ of undercut < 1/32″	2″–3″ of undercut < 1/32″	3″–4″ of undercut < 1/32″	4″–5″ of undercut < 1/32″	5″–7″ of undercut or undercut > 1/32″
Inclusions	0 inclusions	1 inclusion	2 inclusions	3 inclusions	4 inclusions	5 or more inclusions
Craters	0–2	2–4	4–6	6–8	8–10	10–12
Porosity	No porosity	1–2 > 1/32″	3–4 > 1/32″	5–6 > 1/32″	7–8 > 1/32″	+9 > 1/32″
Uniform bead size	± ⅛″	± 3/16″	± ¼″			> ¼″
Overlap	No overlap	1″–2″	2″–3″	3″–4″	4″–5″	+5″
Cleans between passes	All clean	1″–2″ uncleaned	2″–3″ uncleaned	3″–4″ uncleaned	4″–5″ uncleaned	+5″ uncleaned

Student Name: ______________________________
Date: ______________________________ ________ F

Fillet Welds With E6010 or E6011 Electrodes Evaluation

PROCEDURE	CRITERIA	POSSIBLE POINTS	SCORE
Assembly	Weld surfaces on pieces are clean	5	
	Pieces are tight	5	
	Pieces have proper lap	5	
	Subtotal	*15*	
Machine Setup	Proper polarity for the size of the electrode	5	
	Proper grounding	5	
	Proper amperage for the size of the electrode	5	
	Subtotal	*15*	
Weld Beads	Uniform appearance of the weld bead face	5	
	Undercut $\leq 3/32''$	5	
	Inclusions	5	
	Craters and restarts filled to the full cross section of the weld bead	5	
	Porosity	5	
	Uniform weld bead size	5	
	Overlap	5	
	Reinforcement $\leq 1/8''$	5	
	Cleans between passes	5	
	Subtotal	*45*	
General	Tool handling	5	
	Followed directions	5	
	Safe work practices	5	
	Cleaned up	5	
	Subtotal	*20*	
		Student score:	

Total Possible Points = 95

Suggested minimum acceptable score: 67 points or ______

Student's Signature: ______________________________

Teacher's Signature: ______________________________

Student Name:____

Date:____

Fillet Welds With E6010 or E6011 Electrodes Safety Evaluation

Competency	Demonstrated			Date Completed	Teacher Initial*	Student Initial**	Comments
	Yes	No	N/A				
1. Wore proper safety equipment							
2. Always worked in a clean work area							
3. Used tools only as designed							
4. Always removed electrode from holder before hanging it up							
5. Attached work clamp properly							
6. Never laid electrode holder on table							
7. Aware of spark direction while grinding							

*Teacher's initial indicates that the student demonstrated competencies for proper and safe use of this tool.

**Student's initial indicates that the student demonstrated competencies for proper and safe use of this tool.

Student's Signature: ________________________________

Teacher's Signature: ________________________________

Project 2

Fillet Welds With E7018 Electrodes

Job Description

To complete this job successfully, the student should be able to assemble the project according to the given procedures and apply sequential weld beads E7018 electrodes, with minimal variation and defects. This will be done in the 1F, 2F, 3F, and 4F positions.

Materials

- two pieces of 1″ x 1″ x ¼″ angle iron cut 7″ long
- ⅛″ or ³⁄₃₂″ E7018 electrodes

Tools

- wire brush
- chipping hammer
- #11 vise-grips

Equipment

- SMAW welder
- grinder with a wire wheel and a grinding wheel

Safety Precautions

- Use the grinder in a safe manner, without injury to self or others.
- Ensure the equipment is in proper working order.
- Wear clear safety glasses at all times while in the shop.
- Wear appropriate clothing and work shoes at all times while in the shop.
- Wear the appropriate PPE while welding.
- Keep the work area clean and safe.

Procedure

1. Prepare and clean the surface to be welded.
2. Stand the two angles on their ends and lap the corners by ¼″, lining up the legs on the angles.
3. Clamp the angles together.
4. Attach the work clamp (ground) to the work surface or project.
5. Turn on the welder.
6. Set the welder to the proper amperage for the size and type of electrode being used. Start in the middle of the settings for the electrodes listed below.
   - E7018, ³⁄₃₂″, 70–140 amperes
   - E7018, ⅛″, 90–185 amperes
7. Insert the bare metal end of the electrode into the electrode holder (stinger).
8. Tack weld both ends of the angles with the electrode chosen.

Procedure, *cont.*

9. Let the teacher check the fit-up.
10. Welding positions:
    - The flat (1F) position—the angles will be sitting on the table like an X.
    - The horizontal (2F) position—the angle irons will be orientated like a +. It will need to be put in a positioner or tacked to the table.
    - The vertical (3F) position—the angles will be standing on end with an open side facing the welder. It will need to be put in a positioner or tacked to the table.
    - The overhead (4F) position—the angle irons will be orientated like a +. It will need to be put in a positioner over the welders head.
11. In the flat and horizontal positions: if the welder is right-handed, weld left to right. If the welder is left-handed, weld right to left. This can also be used in the overhead position.
12. In the overhead position, the welder can also stand at the end of the angles and weld towards or away from his or her self.
13. In the vertical position, the welder is only allowed to weld from the bottom to the top (Vertical Up) of the angles.
14. Let the teacher check the position of the angles.
15. Use one of the electrode patterns listed in the chapter.
16. To avoid stopping and starting more than necessary, sit or stand in a position that allows the complete range of motion for the work piece.
17. Always remove the electrode from the electrode holder before hanging the holder up.
18. Never lay the electrode holder down on the table—with or without the electrode in it.
19. Using the scratch or tapping method, start the electrode and use the welders preferred technique to weld the root pass.
20. Weld the root bead, keeping it straight and uniform in appearance.
21. Remove any slag with a chipping hammer.
22. Clean the piece with a wire brush or grinder with a wire wheel.
23. The teacher will review the weld before continuing.
24. Remove any slag and clean each weld bead before welding the next weld bead in the sequence.
25. Continue to weld in the sequence shown in the figures on the next page; show the teacher every layer of weld beads for further instruction.
26. Weld stringer beads should be laid half on top of each other; this is achieved by welding along the toe of the stringer bead before.
27. Dip the angle irons in a cooling tank every 2–3 weld layers.
28. When the one side of the angle is completed, let the teacher review the work.
29. Repeat the weld bead sequence on the opposite side if more practice is required.
30. There should be equal amounts of weld on each leg of the angle.

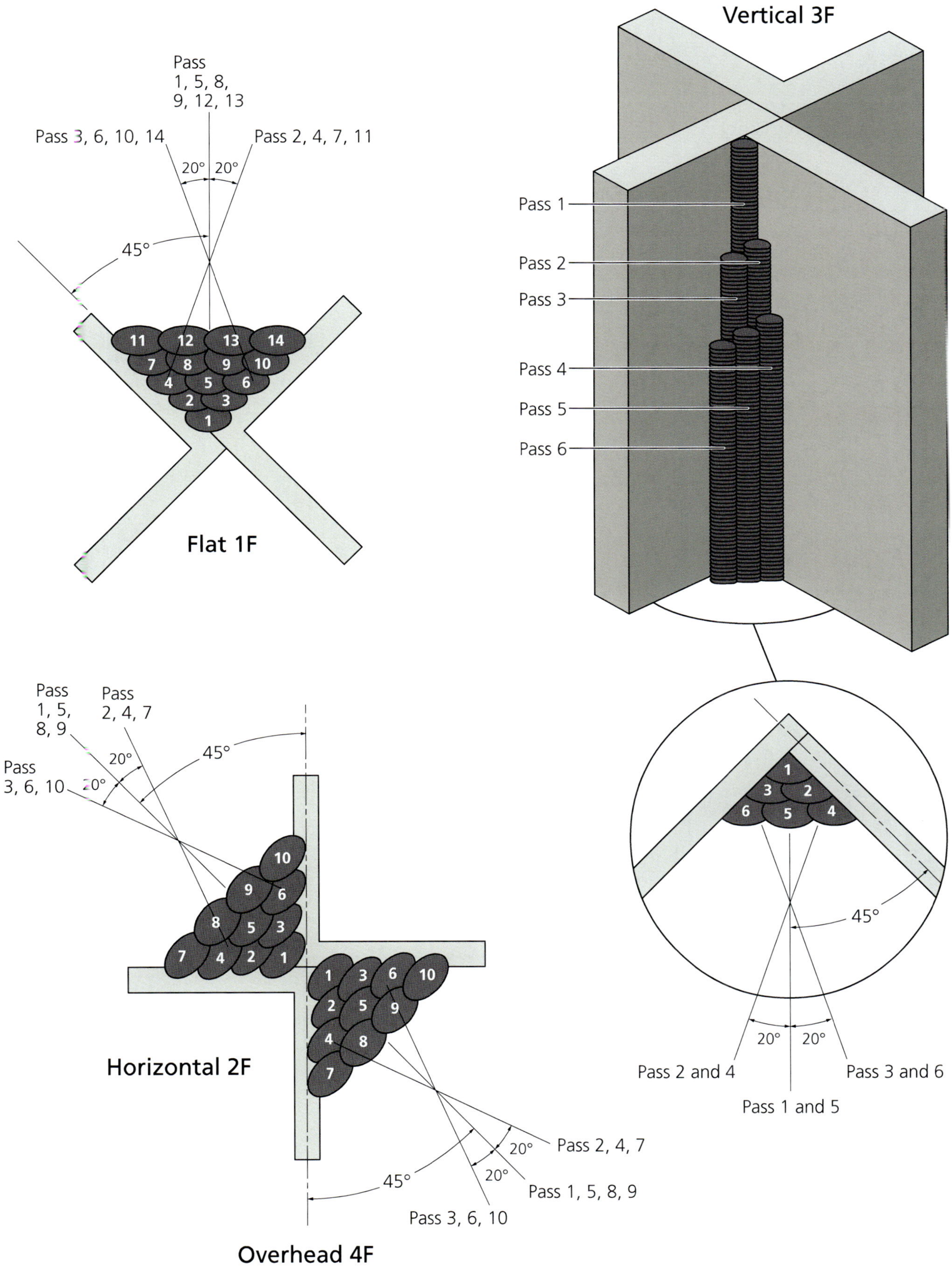
Pass
1, 5, 8,
9, 12, 13
Pass 3, 6, 10, 14
Pass 2, 4, 7, 11
20°
20°
45°
11
12
13
14
7
8
9
10
4
5
6
2
3
1
Flat 1F
Vertical 3F
Pass 1
Pass 2
Pass 3
Pass 4
Pass 5
Pass 6
1
3
2
6
5
4
45°
20°
20°
Pass 2 and 4
Pass 3 and 6
Pass 1 and 5
Pass
1, 5,
8, 9
Pass
2, 4, 7
20°
45°
Pass
3, 6, 10
20°
10
9
6
8
5
3
7
4
2
1
1
3
6
10
2
5
9
4
8
7
Horizontal 2F
Pass 2, 4, 7
20°
20°
Pass 1, 5, 8, 9
45°
Pass 3, 6, 10
Overhead 4F

Fillet Welds With E7018 Electrodes Rubric

POINTS:	5	4	3	2	1	0
Criteria:						
Pieces clean	Completely clean	¾ surfaces clean	½ surfaces clean	¼ surfaces clean	⅛ surfaces clean	Not Cleaned
Pieces tight	Tight	1/16″ gap	⅛″ gap			+⅛″
Pieces lap	¼″ lap	⅜″ lap	½″ lap	⅝″ lap		Assembled wrong
Proper polarity	Correct					Incorrect
Proper grounding	Lead tight to machine and work clamp tight to work piece	Lead tight to machine and work clamp is on coated surface of work piece	Lead tight to machine or work clamp is tight to work piece, not both	Work clamp not on work piece, lead tight to machine		Not grounded
Proper amperage	± 10 amps	± 15 amps	± 20 amps	± 25 amps	± 30 amps	± 35 amps
Appearance of weld bead face	0″–1″ Bad appearance	1″–3″ Bad appearance	3″–5″ Bad appearance	5″–7″ Bad appearance	7″–9″ Bad appearance	+9″ Bad appearance
Undercut	No undercut	1″–2″ of undercut < 1/32″	2″–3″ of undercut < 1/32″	3″–4″ of undercut < 1/32″	4″–5″ of undercut < 1/32″	5″–7″ of undercut or undercut > 1/32″
Inclusions	0 inclusions	1 inclusion	2 inclusions	3 inclusions	4 inclusions	5 or more inclusions
Craters	0–2	2–4	4–6	6–8	8–10	10–12
Porosity	No porosity	1–2 > 1/32″	3–4 > 1/32″	5–6 > 1/32″	7–8 > 1/32″	+9 > 1/32″
Uniform bead size	± ⅛″	± 3/16″	± ¼″			> ¼″
Overlap	No overlap	1″–2″	2″–3″	3″–4″	4″–5″	+5″
Cleans between passes	All clean	1″–2″ uncleaned	2″–3″ uncleaned	3″–4″ uncleaned	4″–5″ uncleaned	+5″ uncleaned

Student Name: ____________________
Date: ____________________

____________ F

Fillet Welds With E7018 Electrodes Evaluation

PROCEDURE	CRITERIA	POSSIBLE POINTS	SCORE
Assembly	Weld surfaces on pieces are clean	5	
	Pieces are tight	5	
	Pieces have proper lap	5	
	Subtotal	*15*	
Machine Setup	Proper polarity for the size of the electrode	5	
	Proper grounding	5	
	Proper amperage for the size of the electrode	5	
	Subtotal	*15*	
Weld Beads	Uniform appearance of the weld bead face	5	
	Undercut ≤ 1/32″	5	
	Inclusions	5	
	Craters and restarts filled to the full cross section of the weld bead	5	
	Porosity	5	
	Uniform weld bead size	5	
	Overlap	5	
	Reinforcement ≤ 1/8″	5	
	Cleans between passes	5	
	Subtotal	*45*	
General	Tool handling	5	
	Followed directions	5	
	Safe work practices	5	
	Cleaned up	5	
	Subtotal	*20*	
		Student score:	

Total Possible Points = 95

Suggested minimum acceptable score: 67 points or ______

Student's Signature: ____________________

Teacher's Signature: ____________________

Student Name:______________________________

Date:______________________________

Fillet Welds With E7018 Electrodes Safety Evaluation

Competency	Demonstrated			Date Completed	Teacher Initial*	Student Initial**	Comments
	Yes	No	N/A				
1. Wore proper safety equipment							
2. Always worked in a clean work area							
3. Used tools only as designed							
4. Always removed electrode from holder before hanging it up							
5. Attached work clamp properly							
6. Never laid electrode holder on table							
7. Aware of spark direction while grinding							

*Teacher's initial indicates that the student demonstrated competencies for proper and safe use of this tool.

**Student's initial indicates that the student demonstrated competencies for proper and safe use of this tool.

Student's Signature: ______________________________

Teacher's Signature: ______________________________

Project 3

Groove Welds With E7018 Electrodes

Job Description

To complete this job successfully, the student should be able to assemble the project according to the given procedures and apply sequential weld beads E7018 electrodes, with minimal variation and defects. This will be done in the 1G, 2G, 3G, and 4G positions.

Materials

- 5 pieces of ⅜″ x 7″ bar stock cut 1 ¾″ with a 22 ½° bevel on each side that are parallel
- 6 pieces of ¼″ x 1″ x 9″ backer
- ⅛″ or 3/32″ E7018 electrodes

Tools

- wire brush
- chipping hammer
- #11 vise-grips

Equipment

- SMAW welder
- grinder with a wire wheel and a grinding wheel

Safety Precautions

- Use the grinder in a safe manner, without injury to self or others.
- Ensure the equipment is in proper working order.
- Wear clear safety glasses at all times while in the shop.
- Wear appropriate clothing and work shoes at all times while in the shop.
- Wear the appropriate PPE while welding.
- Keep the work area clean and safe.

Procedure

1. Prepare and clean the surface to be welded.
2. Place three pieces of the backer on the table. Two will help hold the bevel pieces in place; the third will be the backer for the groove.
3. Place two of the ⅜″ x 7″ pieces on top of the backers, forming a groove on the middle piece of backer.
4. Use one piece of backer to set the root width of ¼″.
5. Clamp the pieces together.
6. Attach the work clamp (ground) to the work surface or project.
7. Turn on the welder.

Procedure, *cont.*

8. Set the welder to the proper amperage for the size and type of electrode being used. Start in the middle of the settings for the electrodes listed below.
   - E7018, 3/32", 70–140 amperes
   - E7018, 1/8", 90–185 amperes
9. Insert the metal end of the electrode into the electrode holder (stinger).
10. Tack both ends of the bevel plates with the electrode chosen.
11. Repeat the process so that two sets of groove plates are made.
12. Place three pieces of backer on the table again. Now place the two groove plates face down and form a groove on the middle piece of backer.
13. Use one piece of backer to set the root width of 1/4".
14. There should now be three grooves fabricated to practice on.
15. Let the teacher check the fit-up.
16. Welding positions:
    - The flat (1G) position—one groove will be sitting table face up on the table.
    - The horizontal (2G) position—the groove will be parallel to the floor, facing forward, and the plate will be placed on its side. It will need to be put in a positioner or tacked to the table.
    - The vertical (3G) position—the plate will be stood on end with the groove facing the welder. It will need to be put in a positioner or tacked to the table.
    - The overhead (4G) position—the groove will be facing the floor and the plate will be parallel to the floor. It will need to be put in a positioner over the welder's head.
17. In the flat and horizontal positions if the welder is right-handed, weld left to right. If the welder is left-handed, weld right to left. This can also be used in the overhead position.
18. In the overhead position, the welder can also stand at the end on the angles and weld toward or away from his or herself.
19. In the vertical position, the welder is only allowed to weld from the bottom to the top (Vertical Up) of the groove.
20. Let the teacher check the position of the plate.
21. Use one of the electrode patterns listed in the chapter.
22. To avoid stopping and starting more than necessary, sit or stand in a position that allows the complete range of motion for the work piece.
23. Always remove the electrode from the electrode holder before hanging it up.
24. Never lay the electrode holder down on the table—with or without the electrode in it.
25. Using the scratch or tapping method start the electrode and use the welder's preferred technique to weld the root pass.
26. Weld the root bead, keeping it straight and uniform in appearance.

27. Remove any slag with a chipping hammer.

28. Clean the piece with a wire brush or grinder with a wire wheel.

29. The teacher will review the weld before continuing.

30. Remove the slag and clean each weld bead before welding the next weld bead in the sequence.

31. Continue to weld in the sequence shown in the figure above; show the teacher every layer of weld beads for further instruction.

32. Weld stringer beads should be laid half on top of each other; this is achieved by welding along the toe of the stringer bead before.

33. Dip the groove plates in a cooling tank after every two to three weld layers.

34. When the one side of the groove plate is completed, let the teacher review the work.

35. Repeat the weld bead sequence on the opposite side if more practice is required.

36. Have good fusion between each bead and the bevels on the plates.

Groove weld practice set up

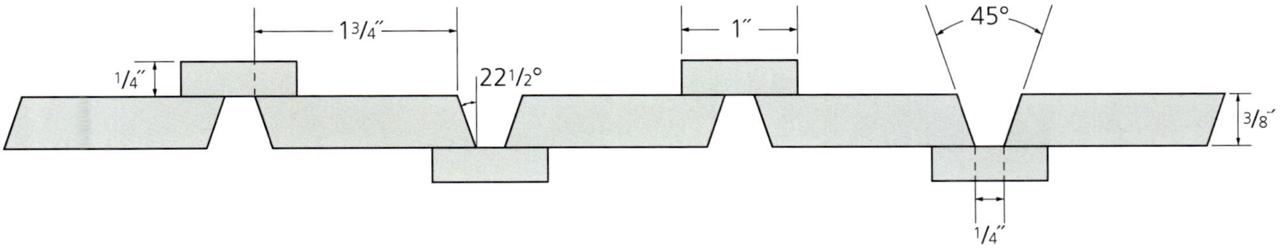

Flat 1G

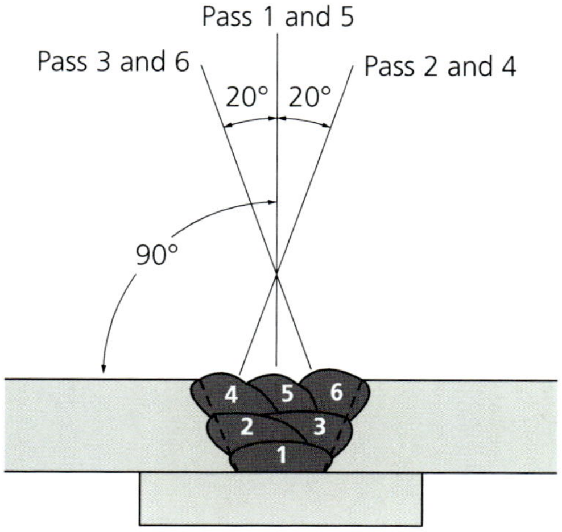

Horizontal 2G

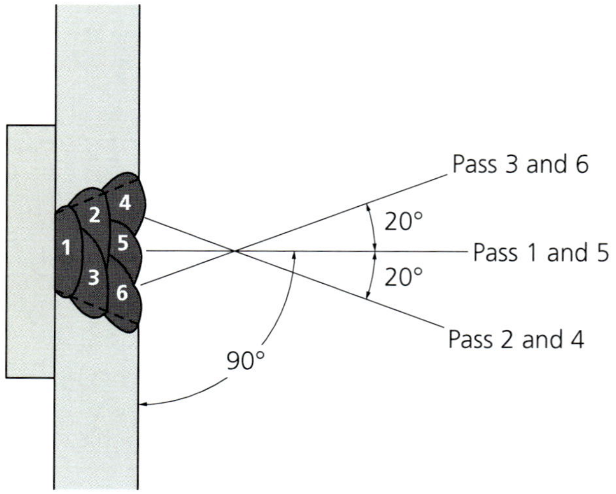

Vertical 3G

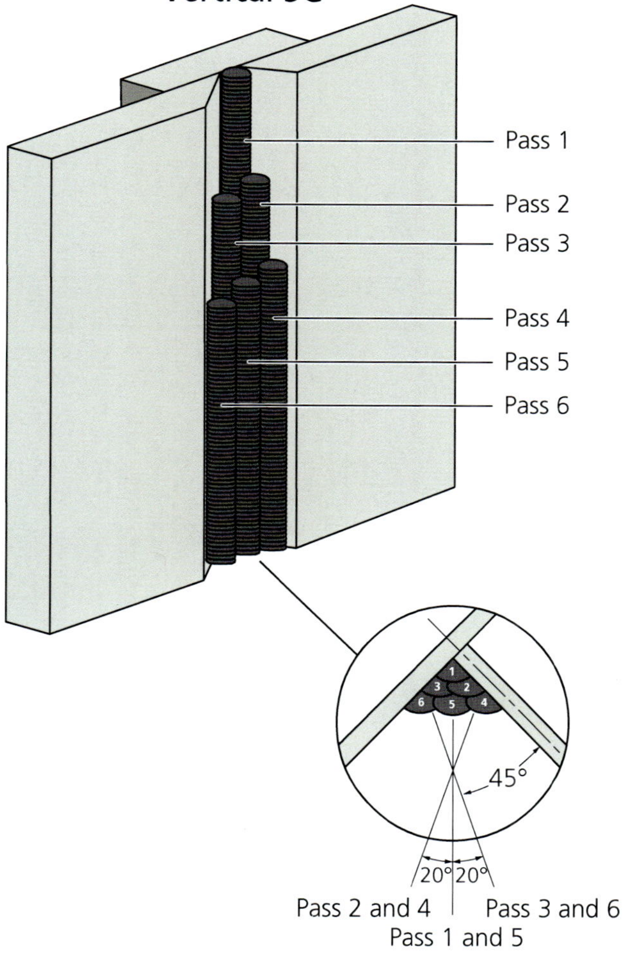

Overhead 4G

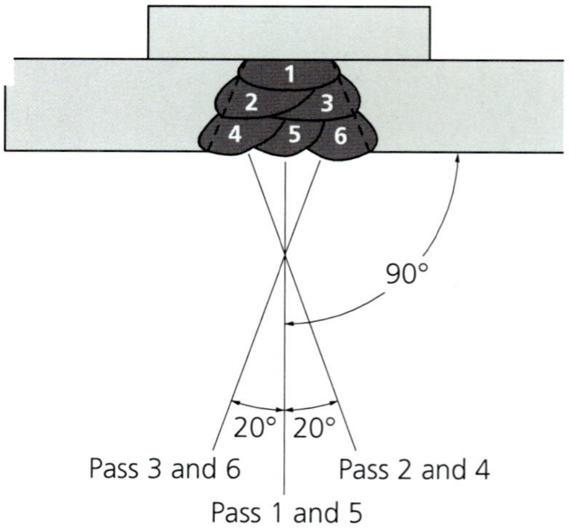

Groove Welds With E7018 Electrodes Rubric

POINTS:	5	4	3	2	1	0
Criteria:						
Pieces clean	Completely clean	$\frac{3}{4}$ surfaces clean	$\frac{1}{2}$ surfaces clean	$\frac{1}{4}$ surfaces clean	$\frac{1}{8}$ surfaces clean	Not Cleaned
Pieces tight	Tight	$\frac{1}{16}''$ gap	$\frac{1}{8}''$ gap			$+\frac{1}{8}''$
Pieces lap	$\frac{1}{4}''$ lap	$\frac{3}{8}''$ lap	$\frac{1}{2}''$ lap	$\frac{5}{8}''$ lap		Assembled wrong
Proper polarity	Correct					Incorrect
Proper grounding	Lead tight to machine and work clamp tight to work piece	Lead tight to machine and work clamp is on coated surface of work piece	Lead tight to machine or work clamp is tight to work piece, not both	Work clamp not on work piece, lead tight to machine		Not grounded
Proper amperage	$\pm$ 10 amps	$\pm$ 15 amps	$\pm$ 20 amps	$\pm$ 25 amps	$\pm$ 30 amps	$\pm$ 35 amps
Appearance of weld bead face	0″–1″ Bad appearance	1″–3″ Bad appearance	3″–5″ Bad appearance	5″–7″ Bad appearance	7″–9″ Bad appearance	+9″ Bad appearance
Undercut	No undercut	1″–2″ of undercut $< \frac{1}{32}''$	2″–3″ of undercut $< \frac{1}{32}''$	3″–4″ of undercut $< \frac{1}{32}''$	4″–5″ of undercut $< \frac{1}{32}''$	5″–7″ of undercut or undercut $> \frac{1}{32}''$
Inclusions	0 inclusions	1 inclusion	2 inclusions	3 inclusions	4 inclusions	5 or more inclusions
Craters	0–2	2–4	4–6	6–8	8–10	10–12
Porosity	No porosity	1–2 $> \frac{1}{32}''$	3–4 $> \frac{1}{32}''$	5–6 $> \frac{1}{32}''$	7–8 $> \frac{1}{32}''$	+9 $> \frac{1}{32}''$
Uniform bead size	$\pm\frac{1}{8}''$	$\pm\frac{3}{16}''$	$\pm\frac{1}{4}''$			$> \frac{1}{4}''$
Overlap	No overlap	1″–2″	2″–3″	3″–4″	4″–5″	+5″
Cleans between passes	All clean	1″–2″ uncleaned	2″–3″ uncleaned	3″–4″ uncleaned	4″–5″ uncleaned	+5″ uncleaned

Student Name:____

Date:____

Groove Welds With E7018 Electrodes Evaluation

PROCEDURE	CRITERIA	POSSIBLE POINTS	SCORE
Assembly	Weld surfaces on pieces are clean	5	
	Pieces are tight to the backer	5	
	Pieces have 1/4″ root gap	5	
	Subtotal	*15*	
Machine Setup	Proper polarity for the size of the electrode	5	
	Proper grounding	5	
	Proper amperage for the size of the electrode	5	
	Subtotal	*15*	
Weld Beads	Uniform appearance of the weld bead face	5	
	Undercut ≤ 1/32″	5	
	Inclusions	5	
	Craters and restarts filled to the full cross section of the weld bead	5	
	Porosity	5	
	Uniform weld bead size	5	
	Overlap	5	
	Reinforcement ≤ 1/8″	5	
	Cleans between passes	5	
	Subtotal	*45*	
General	Tool handling	5	
	Followed directions	5	
	Safe work practices	5	
	Cleaned up	5	
	Subtotal	*20*	

Total Possible Points = 95

Student score: ______

Suggested minimum acceptable score: 67 points or ______

Student's Signature: ____

Teacher's Signature: ____

Student Name:____________________

Date:____________________

Groove Welds With E7018 Electrodes Safety Evaluation

Competency	Demonstrated			Date Completed	Teacher Initial*	Student Initial**	Comments
	Yes	No	N/A				
1. Wore proper safety equipment							
2. Always worked in a clean work area							
3. Used tools only as designed							
4. Always removed electrode from holder before hanging it up							
5. Attached work clamp properly							
6. Never laid electrode holder on table							
7. Aware of spark direction while grinding							

*Teacher's initial indicates that the student demonstrated competencies for proper and safe use of this tool.

**Student's initial indicates that the student demonstrated competencies for proper and safe use of this tool.

Student's Signature: ____________________

Teacher's Signature: ____________________

Project 4

E7018 2G Weld Test

Job Description

To complete this job successfully, the student should be able to assemble the project according to the given procedures and apply sequential weld beads E7018 electrodes, with minimal variation and defects. This will be done in the 2G position. The weld will be evaluated to AWS D1.1-2015 Structural Welding Code-Steel.

Materials

- 2 pieces of ⅜″ x 7″ bar stock cut 3 ¼″ with a 22 ½° bevel on each side that are parallel
- 4 pieces of ¼″ x 1″ x 9″ backer
- ⅛″ or ³⁄₃₂″ E7018 electrodes

Tools

- wire brush
- chipping hammer
- #11 vise-grips

Equipment

- SMAW welder
- grinder with a wire wheel and a grinding wheel

Safety Precautions

- Use the grinder in a safe manner, without injury to self or others.
- Ensure the equipment is in proper working order.
- Wear clear safety glasses at all times while in the shop.
- Wear appropriate clothing and work shoes at all times while in the shop.
- Wear the appropriate PPE while welding.
- Keep the work area clean and safe.

Procedure

1. Prepare and clean the surface to be welded.
2. Place three pieces of the backer on the table. Two will help hold the bevel pieces in place; the third will be the backer for the groove.
3. Place two of the ⅜″ x 7″ pieces on top of the backers, forming a groove on the middle piece of backer.
4. Use one piece of backer to set the root width of ¼″
5. Clamp the pieces together.
6. Attach the work clamp (ground) to the work surface or project.
7. Turn on the welder.

8. Set the welder to the proper amperage for the size and type of electrode being used. Start in the middle of the settings for the electrodes listed below.
   - E7018, 3⁄32″, 70–140 amperes
   - E7018, 1⁄8″, 90–185 amperes
9. Insert the metal end of the electrode into the electrode holder (stinger).
10. Tack both ends of the bevel plates with the electrode chosen.
11. The teacher will check the fit-up of the test plates.
12. Welding position:
   - The horizontal (2G) position—the groove will be parallel to the floor, facing forward, and the plate will be placed on its side; it will need to be put in a positioner or tacked to the table. See the following page.
13. The teacher will check the plate for correct position. The plate cannot be moved out of position once started.
14. In the horizontal position, if the welder is right-handed, weld left to right. If the welder is left-handed, weld right to left. This can also be used in the overhead position.
15. Use one of the electrode patterns listed in the chapter.
16. To avoid stopping and starting more than necessary, sit or stand in a position that allows the complete range of motion for the work piece.
17. Always remove the electrode from the electrode holder before hanging it up.
18. Never lay the electrode holder down on the table—with or without the electrode in it.
19. Using the scratch or tapping method, start the electrode and use the welders preferred technique to weld the root pass.
20. Weld the root bead, keeping it straight and uniform in appearance.
21. Remove any slag with a chipping hammer.
22. Clean the piece with a wire brush or grinder with a wire wheel.
23. The teacher will review the root bead weld before continuing. If the root bead passes visual inspection, the student may continue.
24. Remove the slag and clean each weld bead before welding the next weld bead in the sequence.
25. The weld stringer beads should be laid half on top of each other; this is achieved by welding along the toe of the stringer bead before.
26. Do *not* dip the test plate in the cooling tank.
27. When the groove is completed, let the teacher review the work before moving the plate.
28. If the test plate passes the cap visual inspection, the teacher will then cool and test the plate.

Note: Base metal = carbon steel plate

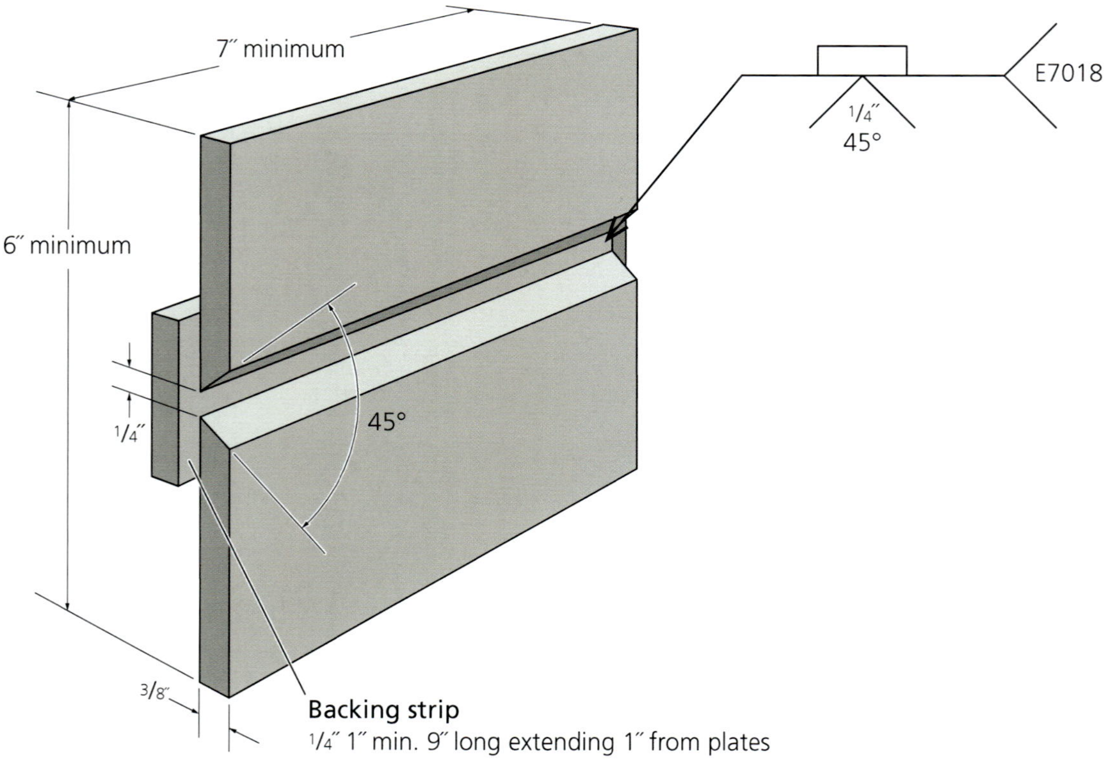

Student Name: ______________________________

Date: ______________________________

E7018 2G Weld Test Evaluation

PROCEDURE	CRITERIA	PASS	FAIL
Assembly	Weld surfaces on pieces are clean		
	Pieces are tight to backer		
	Pieces have ¼″ root gap		
	Overall asembly		
Machine Setup	Proper polarity for the electrode		
	Proper amperage for the size of the electrode		
	Overall machine setup		
Visual Inspection Root	Any crack shall be unacceptable		
	The root shall have complete fusion between adjacent layers of weld metal and between weld metal and base metal. (No cold lap, slag inclusions, or overlap)		
	Shall not have inadequate joint penetration. (No slag trapped between backer and weld metal, cut edge of plate not visible)		
	Maximum bead thickness ¼″		
	Overall root visual inspection results		
Visual Inspection Cap	Any crack shall be unacceptable		
	All craters shall be filled to the full cross section of the weld. (Any under fill shall be unacceptable)		
	Weld reinforcement shall not exceed ⅛″		
	The weld profile shall have complete fusion between adjacent layers of weld metal and between weld metal and base metal. (No-cold lap, slag inclusions, or overlap)		
	Undercut shall not exceed 1/32″		
	Maximum bead thickness ¼″		
	Overall cap visual inspection results		
Bend Tests	Shall contain no discontinuities exceeding ¼″ in any direction		
	⅜″ the sum of the greatest dimension of all discontinuities exceeding 1/32″, but less than or equal to ⅛″		
	¼″ the maximum corner crack, except when that corner crack results from visible slag inclusion or other fusion type discontinuity, then the ⅛″ maximum shall apply		
	Overall bend tests results		
Weld Test Results E7018 2G			

Student's Signature: ______________________________

Teacher's Signature: ______________________________

Student Name: ______________________________

Date: ______________________________

E7018 2G Weld Test Safety Evaluation

Competency	Demonstrated			Date Completed	Teacher Initial*	Student Initial**	Comments
	Yes	No	N/A				
1. Wore proper safety equipment							
2. Always worked in a clean work area							
3. Used tools only as designed							
4. Always removed electrode from holder before hanging it up							
5. Attached work clamp properly							
6. Never laid electrode holder on table							
7. Aware of spark direction while grinding							

*Teacher's initial indicates that the student demonstrated competencies for proper and safe use of this tool.

**Student's initial indicates that the student demonstrated competencies for proper and safe use of this tool.

Student's Signature: ______________________________

Teacher's Signature: ______________________________

Project 5

E7018 3G Weld Test

Job Description

To complete this job successfully, the student should be able to assemble the project according to the given procedures and apply sequential weld beads E7018 electrodes, with minimal variation and defects. This will be done in the 3G position. The weld will be evaluated to AWS D1.1-2015 Structural Welding Code-Steel.

Materials

- 2 pieces of ⅜″ x 7″ bar stock cut 3 ¼″ with a 22 ½° bevel on each side that are parallel
- 4 pieces of ¼″ x 1″ x 9″ backer
- ⅛″ or 3⁄32″ E7018 electrodes

Tools

- wire brush
- chipping hammer
- #11 vise-grips

Equipment

- SMAW welder
- grinder with a wire wheel and a grinding wheel

Safety Precautions

- Use the grinder in a safe manner, without injury to self or others.
- Ensure the equipment is in proper working order.
- Wear clear safety glasses at all times while in the shop.
- Wear appropriate clothing and work shoes at all times while in the shop.
- Wear the appropriate PPE while welding.
- Keep the work area clean and safe.

Procedure

1. Prepare and clean the surface to be welded.
2. Place three pieces of the backer on the table. Two will help hold the bevel pieces in place, the third will be the backer for the groove.
3. Place two of the ⅜″ x 7″ pieces on top of the backers, forming a groove on the middle piece of backer.
4. Use one piece of backer to set the root width of ¼‴.
5. Clamp the pieces together.
6. Attach the work clamp (ground) to the work surface or project.
7. Turn on the welder.

Procedure, *cont.*

8. Set the welder to the proper amperage for the size and type of electrode being used. Start in the middle of the settings for the electrodes listed below.
   - E7018, 3⁄32″, 70–140 amperes
   - E7018, 1⁄8″, 90–185 amperes
9. Insert the metal end of the electrode into the electrode holder (stinger).
10. Tack both ends of the bevel plates with the electrode chosen.
11. The teacher will check the fit-up of the test plates.
12. Welding position:
    - The horizontal (3G) position—the plate will be stood on end with the groove facing the welder. It will need to be put in a positioner or tacked to the table. See the next page.
13. The teacher will check the plate for correct position. The plate cannot be moved out of position once started.
14. In the horizontal position, if the welder is right-handed, weld left to right. If the welder is left-handed, weld right to left. This can also be used in the overhead position.
15. Use one of the electrode patterns listed in the chapter.
16. To avoid stopping and starting more than necessary, sit or stand in a position that allows the complete range of motion for the work piece.
17. Always remove the electrode from the electrode holder before hanging it up.
18. Never lay the electrode holder down on the table—with or without the electrode in it.
19. Using the scratch or tapping method, start the electrode and use the welders preferred technique to weld the root pass.
20. Weld the root bead, keeping it straight and uniform in appearance.
21. Remove any slag with a chipping hammer.
22. Clean the piece with a wire brush or grinder with a wire wheel.
23. The teacher will review the root bead weld before continuing. If the root bead passes visual inspection, the student may continue.
24. Remove the slag and clean each weld bead before welding the next weld bead in the sequence.
25. The weld stringer beads should be laid half on top of each other; this is achieved by welding along the toe of the stringer bead before.
26. Do *not* dip the test plate in the cooling tank.
27. When the groove is completed, let the teacher review the work before moving the plate.
28. If the test plate passes the cap visual inspection, the teacher will then cool and test the plate.

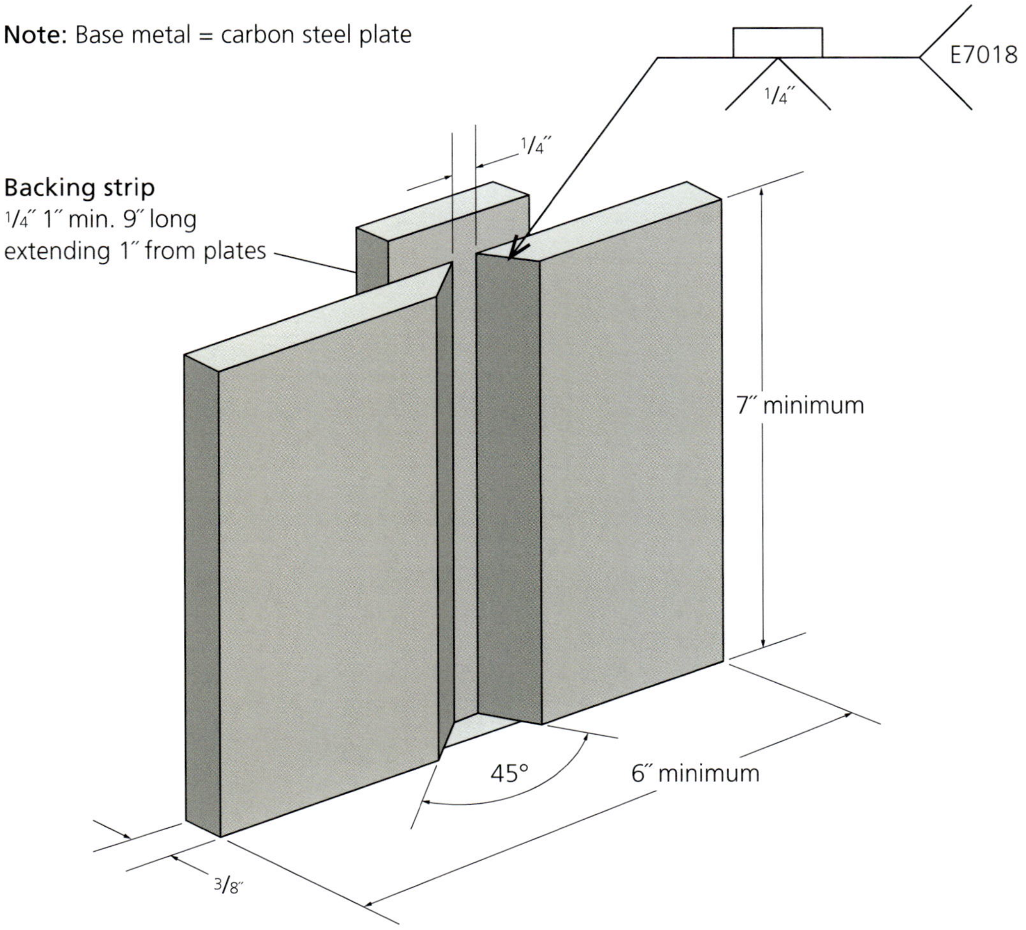
Note: Base metal = carbon steel plate
E7018
1/4″
1/4″
Backing strip
1/4″ 1″ min. 9″ long
extending 1″ from plates
7″ minimum
45°
6″ minimum
3/8″

Student Name: ____________________

Date: ____________________

E7018 3G Weld Test Evaluation

PROCEDURE	CRITERIA	PASS	FAIL
Assembly	Weld surfaces on pieces are clean		
	Pieces are tight to backer		
	Pieces have ¼″ root gap		
	Overall asembly		
Machine Setup	Proper polarity for the electrode		
	Proper amperage for the size of the electrode		
	Overall machine setup		
Visual Inspection Root	Any crack shall be unacceptable		
	The root shall have complete fusion between adjacent layers of weld metal and between weld metal and base metal. (No cold lap, slag inclusions, or overlap)		
	Shall not have inadequate joint penetration. (No slag trapped between backer and weld metal, cut edge of plate not visible)		
	Maximum bead thickness ¼″		
	Overall root visual inspection results		
Visual Inspection Cap	Any crack shall be unacceptable		
	All craters shall be filled to the full cross section of the weld. (Any under fill shall be unacceptable)		
	Weld reinforcement shall not exceed ⅛″		
	The weld profile shall have complete fusion between adjacent layers of weld metal and between weld metal and base metal. (No-cold lap, slag inclusions, or overlap)		
	Undercut shall not exceed 1/32″		
	Maximum bead thickness ¼″		
	Overall cap visual inspection results		
Bend Tests	Shall contain no discontinuities exceeding ¼″ in any direction		
	⅜″ the sum of the greatest dimension of all discontinuities exceeding 1/32″, but less than or equal to ⅛″		
	¼″ the maximum corner crack, except when that corner crack results from visible slag inclusion or other fusion type discontinuity, then the ⅛″ maximum shall apply		
	Overall bend tests results		
Weld Test Results E7018 3G			

Student's Signature: ____________________

Teacher's Signature: ____________________

Student Name: ____________________

Date: ____________________

E7018 3G Weld Test Safety Evaluation

Competency	Demonstrated			Date Completed	Teacher Initial*	Student Initial**	Comments
	Yes	No	N/A				
1. Wore proper safety equipment							
2. Always worked in a clean work area							
3. Used tools only as designed							
4. Always removed electrode from holder before hanging it up							
5. Attached work clamp properly							
6. Never laid electrode holder on table							
7. Aware of spark direction while grinding							

*Teacher's initial indicates that the student demonstrated competencies for proper and safe use of this tool.

**Student's initial indicates that the student demonstrated competencies for proper and safe use of this tool.

Student's Signature: ____________________

Teacher's Signature: ____________________

CHAPTER 6

Weld Symbols

Weld symbols are used to give instructions. These symbols appear on engineering drawings and tell a welder exactly what to weld, where to weld, what type of welds to make, and what welding process to use, among other things. Some symbols are required, that is, they are used on every drawing. Other symbols are optional and whether they appear or not is dependent upon the engineer or the project requirements. The weld symbols shown in this chapter are drawn in accordance with AWS standards. Welding organizations and groups that have developed welding standards will also be introduced in this chapter.

Knowing what the symbols are and how to read them will result in better communication and welding quality.

Contents

Objectives

- Describe the basic weld symbols.
- Describe the optional symbols.
- List the titles of those organizations that develop welding standards.

Key Terms

American Society of Mechanical Engineers (ASME) organization that sets internationally recognized industrial and manufacturing codes and standards that enhance safety

American Welding Society (AWS) organization that is completely devoted to welding standards

ASTM International organization formerly known as American Society for Testing and Materials; develops welding standards and publications among its other services

arrow indicates the part of a reference line applying to the nearest joint while information on the other side of the arrow applies to the farthest joint; a required weld symbol

back weld weld made at the back of a single groove weld

backgouging removing weld metal and base metal from the weld root side of a welded joint; eases complete fusion and complete joint penetration upon further welding from the weld root side

backing box that sits on top of the reference line and indicates that a backing strip or bar should be used when making the weld

consumable insert empty box affixed to the reference line opposite a groove-weld symbol

contour symbol line drawn above the reference line that indicates the shape of the finished weld

Continued on next page.

Key Terms

Key Terms, *cont.*

field weld flag that juts away from the reference line, indicating that a weld should be made at the installation point or the place of construction, not at the shop

melt-through solid half circle attached to the reference line, indicating complete joint penetration with root reinforcement in welds made from one side

spacer rectangle that sits within the reference line and indicates that a joint spacer should be used

reference line information about the welding process is located on this line; a required weld symbol

tail used only when there is additional information supplementing the required instructions found on the reference line

weld-all-around circle around the junction of the arrow and the reference line

weld symbols lines formed in a certain way that indicates welding actions that should take place

Key Terms (continued

1 Weld Symbols

When welding instructions need to be given on a drawing, engineers and foremen use **weld symbols** to communicate. There are two classes of symbols: required and optional.

Required symbols are those that must appear on the drawing. These symbols consist of the reference line and the arrow. Additional information that may be provided but is not required uses optional symbols. Optional information includes the dimensions and other data, finishing symbols, specification process or other references, and supplementary symbols.

Symbols are placed in a standard order. Each element is represented in relation to the other symbols. Figure 1 shows the standard locations of various symbols.

Figure 1
Standard weld symbol locations

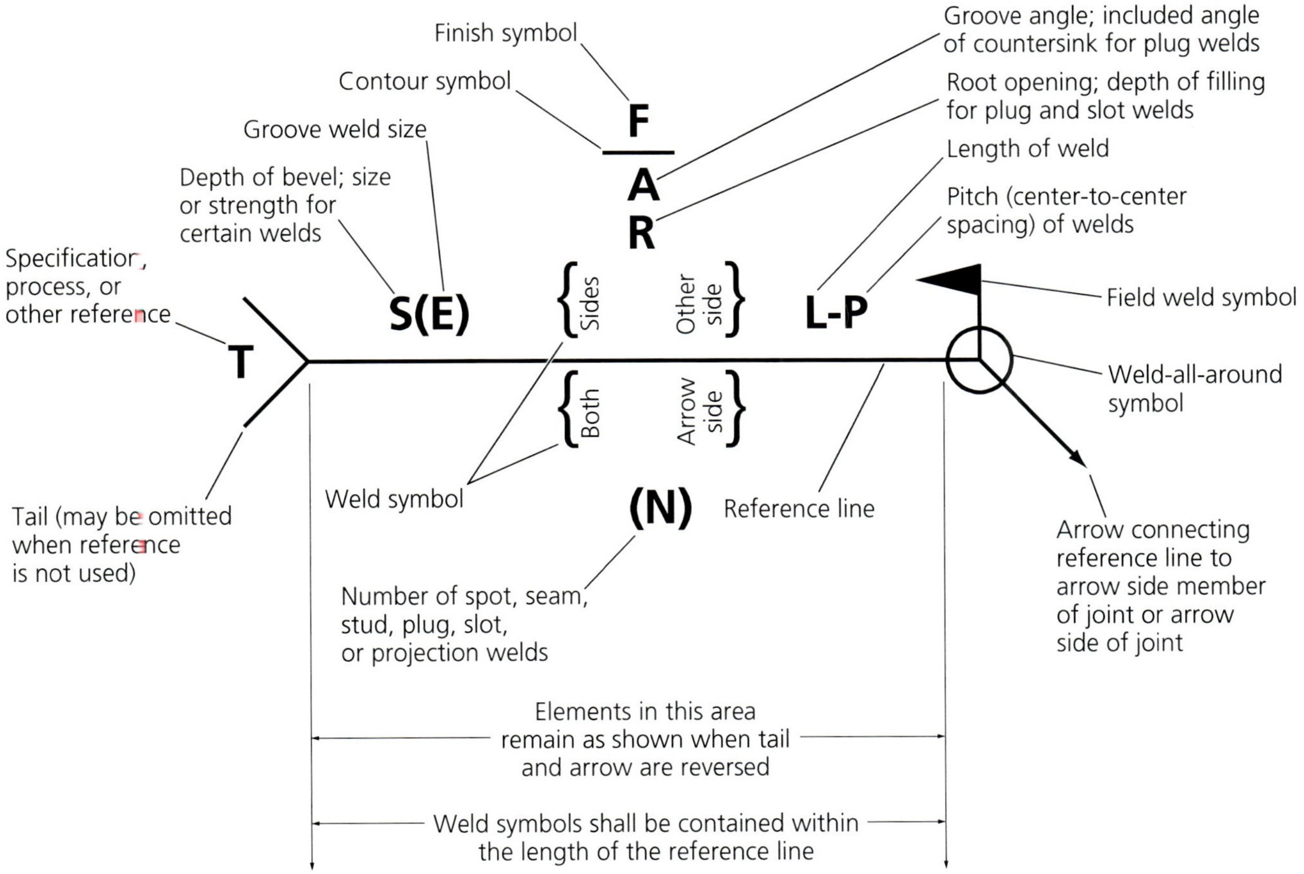

2 Required Symbols

There are only two required weld symbols: the reference line and the arrow. From these, the basic welding instructions are given.

Reference Line

Information about the welding job is located on the **reference line**. Any other symbols used in the welding instructions are located on or around this line. Multiple reference lines can be used with a single arrow to indicate a sequence of welding events. Reference lines are read starting with the closest line to the arrow tip then outward. They are read from the bottom up.

Reference lines may also be used to give supplemental weld information and inspection requirements. Figure 2 shows multiple reference lines and

Figure 2
Multiple reference lines

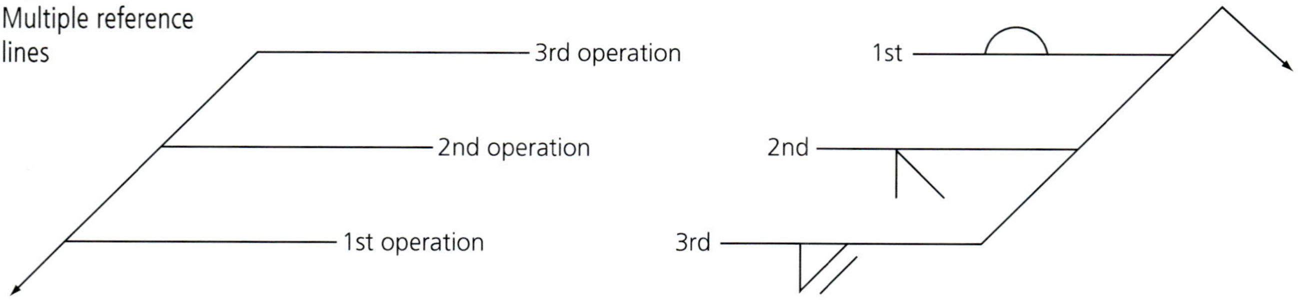

Figure 3
Using multiple reference lines

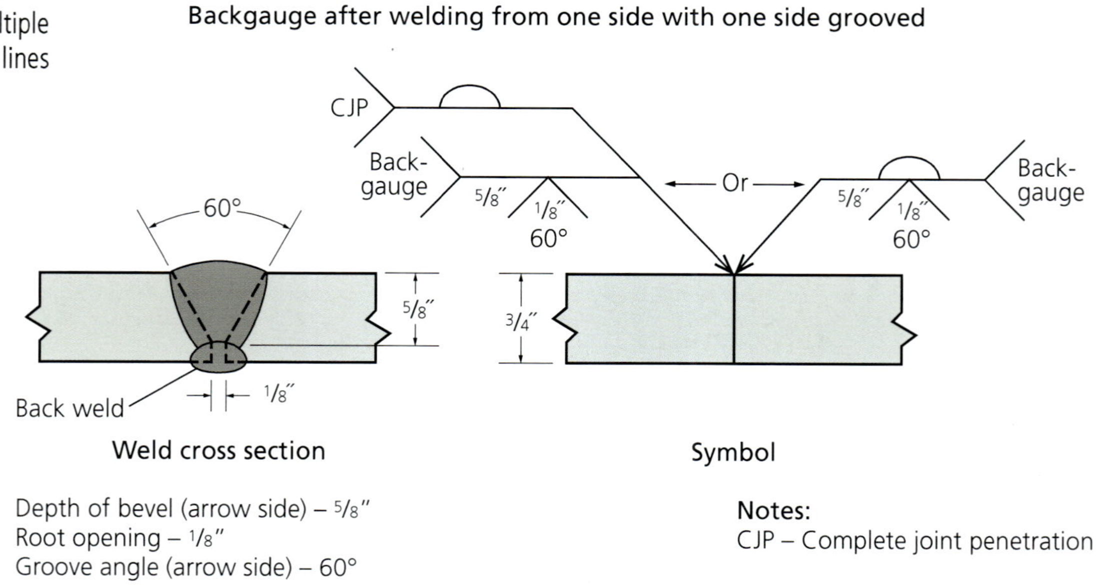

the order in which the instructions should be read and followed. Figure 3 shows a weld symbol.

Arrow

There are only two required parts to a welding symbol: a reference line and an arrow. See Figure 4. The reference line contains information about the welding job. It is always drawn horizontally. Used to apply weld symbols and other data, it has a particular significance that stays the same regardless of any elements that are added to it. The lower side of the reference line is termed the arrow side and the upper side of the reference line is termed the other side.

> ...There are only two required weld symbols: the reference line and the arrow.

Figure 4
Reading the arrow symbol

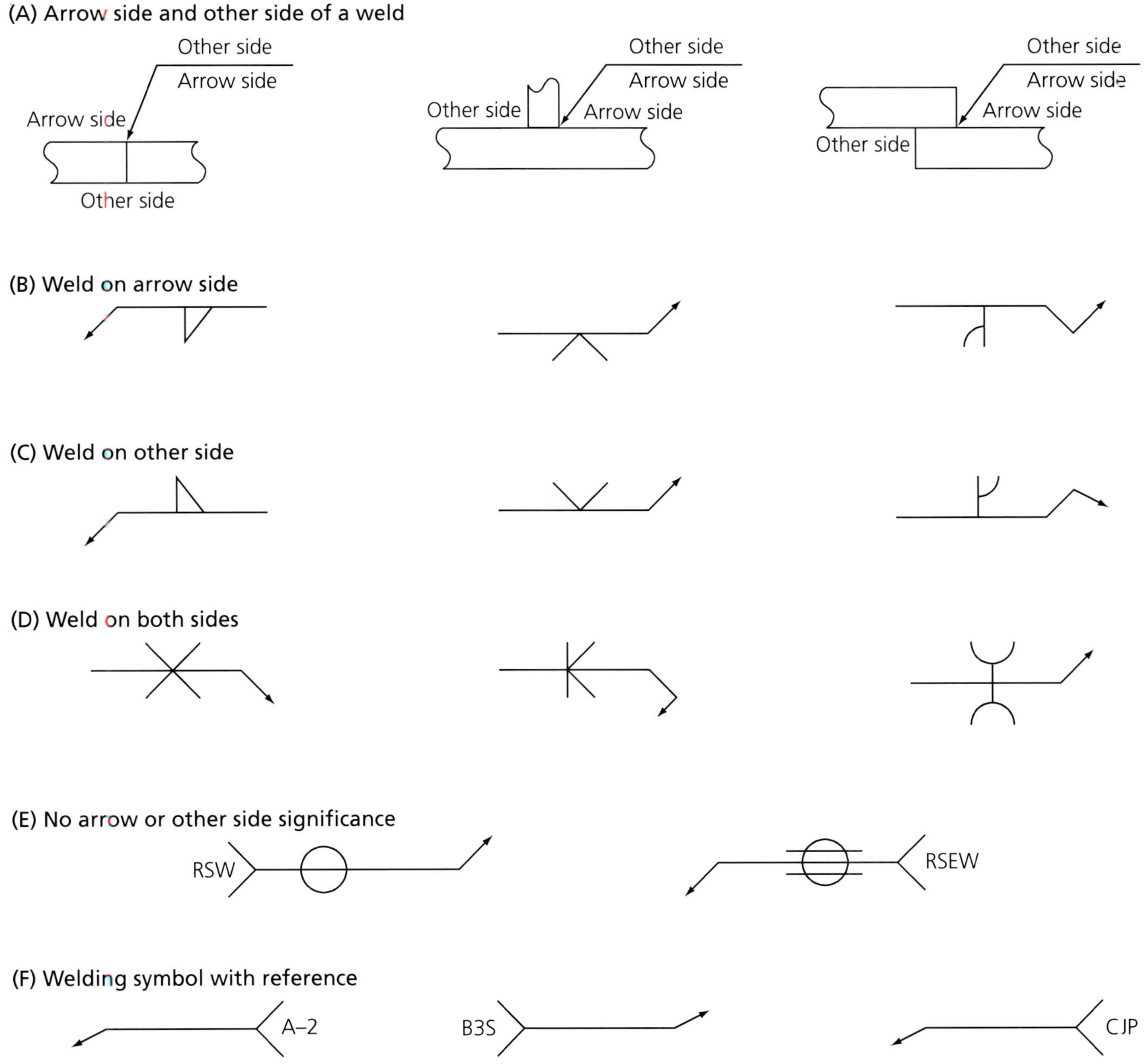

The **arrow** simply connects the reference line to the weld joint or the area to be welded. If the other side of the joint is to be welded, the weld symbol is placed above the reference line. If the arrow side of the joint is to be welded, the weld symbol is placed below the reference line. If both sides of the joint need to be welded, weld symbols are placed above and below the reference line.

Tail

When there are additional welding instructions or information that needs to be shown, that information is placed in the **tail**. The tail is only used if the information included is supplemental to the required instructions found on the reference line. Examples of what this information could be are the welding process, process variation, method of application, and welding procedure number.

There are 16 basic weld symbols. Each of these symbols forms the building blocks of most welding projects. Figure 5 illustrates the symbols.

Self Check

1. Is one weld symbol more important than another?
2. Can very complicated instructions be simplified using weld symbols?

Figure 5
Weld symbols

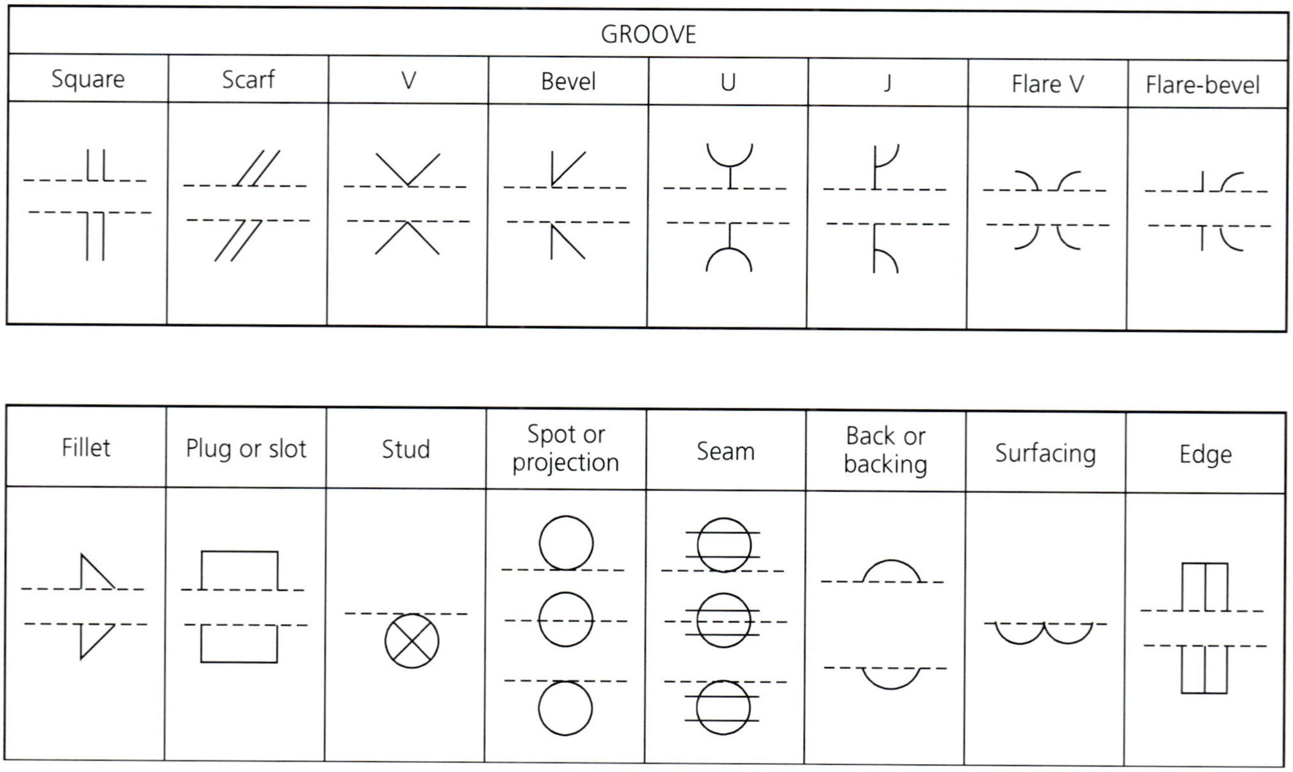

3 Optional Symbols

Optional, or additional, symbols are used to provide more information, instructions, or requirements. They are only used in addition to the required symbols. Figure 6 shows supplemental weld symbols. There are six supplemental symbols and each is explained below. However, there are a number of other optional weld symbols used for specific situations and techniques.

Weld-All-Around Symbol

The **weld-all-around** symbol is a circle around the junction of the arrow and the reference line. It always means to weld all around the area indicated by the arrow. If there are additional welding instructions, it may mean to weld all around different parts of the object to be welded.

Field Weld Symbol

The **field weld** symbol is a flag that juts away from the reference line in either direction. The field weld symbol indicates that a weld should be made at the installation point or the place of construction and not in a shop.

Melt-Through Symbol

The **melt-through** symbol is a solid half-circle attached to the reference line. It calls for complete joint penetration with root reinforcement in welds made from one side. It is usually placed on the other side of the reference line. The placement indicates where the reinforcement is requested.

Figure 6
Supplemental weld symbols

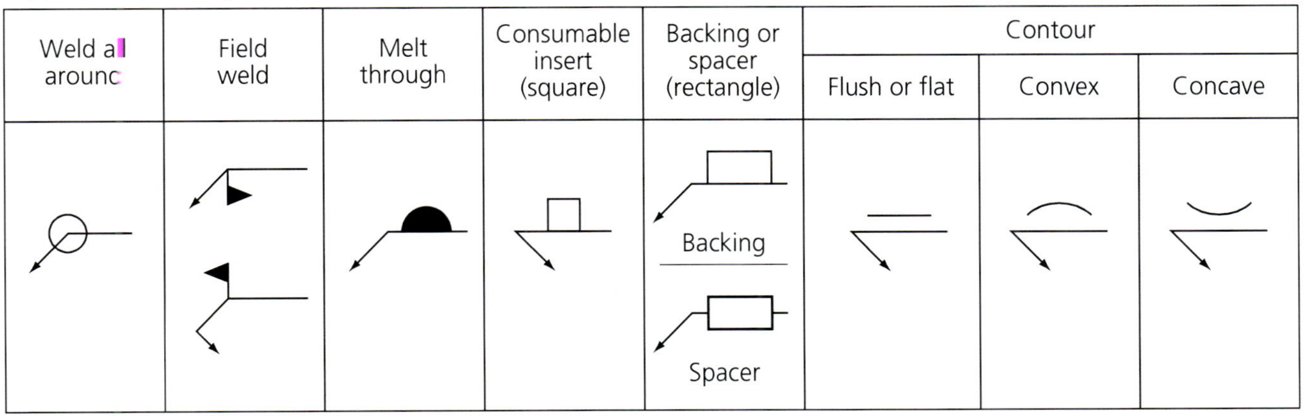

Consumable Insert Symbol

The **consumable insert** symbol is an empty box affixed to the reference line. It is placed on the reference line opposite the *groove-weld symbol*. The specific information on the consumable is placed in the tail of the weld symbol.

Backing and Spacer Symbols

The backing and spacer symbols are slight variations on an empty rectangle. The **backing** symbol is a box that sits on top of the reference line, while the **spacer** symbol is a rectangle that sits within the reference line. The backing symbol is used to indicate that a backing strip or bar should be used to make the weld. The backing symbol is always used in combination with a groove-weld symbol so as to not confuse it with the *plug-* or *slot-weld* symbol. The tail of the weld symbol contains the backing type, material, and dimensions of the backing plate.

The spacer symbol is used to indicate a joint spacer should be used. The material and dimensions of the joint spaces are indicated in the weld symbol tail or elsewhere in the drawing.

Contour Symbol

The **contour symbol** is a line drawn above the reference line that indicates the shape of the finished weld. The line varies in shape from flush, or flat, to *convex* or *concave*. How the shape is to be achieved is often indicated above the symbol. The finishing methods are chipping (C), grinding (G), machining (M), or unspecified (U).

4 Developing Welding Standards

With the onset of World War I, welding became a vital part of the manufacturing industry. Today, there are over ninety different welding processes in use. As technology began to develop, so did the need for standards. Several important organizations were formed over the last century to improve safety and work standards. The welding industry adheres to many of the standards established by these groups.

ASTM International (ASTM)

ASTM International, known until 2001 as the American Society for Testing and Materials (ASTM), formed in 1898, and was created because the rapidly growing railroad network was experiencing frequent rail breaks. A group of engineers and scientists got together to address the problem and formulate a solution. As a result of their work, steel for rail construction became standardized and railroad transportation became safer and more reliable. ASTM is known for its best-in-class practices for standards development and delivery across a wide range of manufacturing and service industries.

American Welding Society (AWS)

In 1919, the **American Welding Society** (**AWS**) was founded as a multifaceted, nonprofit organization with the goal of advancing the science, technology, and application of welding and related joining disciplines. AWS supports the education and technological advances for the welding industry.

American Society of Mechanical Engineers (ASME)

Founded in 1880, the **American Society of Mechanical Engineers** (**ASME**) now has over 120,000 members worldwide. It is focused on technical, educational, and research issues of the engineering and technology community. ASME sets internationally recognized industrial and manufacturing codes and standards that enhance safety.

Self Check

1. Are the optional symbols listed above the only other optional symbols used?
2. What is the weld symbol for backing?

Summary

Weld symbols are used to communicate welding instructions. There are required basic symbols that must appear on engineering drawings, for instance, reference line, arrow, and sometimes a tail, and optional symbols that are used to convey additional information as necessary. Knowing the symbols and being able to understand the information and directions they convey will make it easier to work efficiently on the jobsite. Organizations have been created to improve welding standards and safety. These organizations are the ASTM, AWS, and ASME.

Chapter Review

Show your understanding of the information in this chapter by answering the questions and filling in the blanks below.

1. There are ____________________ required weld symbols.
2. There are ____________________ basic weld symbols.
3. Weld symbols are used to give ____________________ to the welder.
4. Reference lines are read from ____________________ to ____________________.
5. Additional information can be given in the ____________________ that is attached to the reference line.
6. The field weld symbol is a:
   a. box
   b. circle
   c. line
   d. flag
7. The melt-through symbol is a:
   a. rectangle
   b. half-circle
   c. square
   d. flag
8. The weld-all-around symbol is a:
   a. box
   b. circle
   c. line
   d. flag
9. The consumable insert symbol is a(n):
   a. empty box above the reference line
   b. circle above the reference line
   c. filled box above the reference line
   d. filled circle through the reference line
10. It is faster to use symbols than to write out instructions. (True; False)

glossary

acetylene popular fuel gas for cutting

alternating current (AC) electric current that reverses direction at regular intervals

American Society of Mechanical Engineers (ASME) organization that sets internationally recognized industrial and manufacturing codes and standards that enhance safety

American Welding Society (AWS) organization that is completely devoted to welding standards

arc blow wandering of the arc stream due to fluctuation of the magnetic field

arc welding various welding processes that use an electric arc from a metal rod to the workpiece to create a weld

arrow indicates the part of a reference line applying to the nearest joint, while information on the other side of the arrow applies to the farthest joint; a required weld symbol

ASTM International organization formerly known as American Society for Testing and Materials; develops welding standards and publications, among its other services

back weld weld made at the back of a single groove weld

backgouging removing weld metal and base metal from the weld root side of a welded joint; eases complete fusion and complete joint penetration upon further welding from the weld root side

backing box that sits on top of the reference line and indicates that a backing strip or bar should be used when making the weld

backing plate used as support when creating a joint weld

bead or **weld bead** continuous deposit of fused metal

bib part of a leather ensemble to protect the upper torso

brazing uniting metal objects at high temperatures by applying any of various nonferrous solders

British thermal unit (Btu) a measure of heat production

cables or leads insulated electrical conductor composed of thousands of flexible thin copper wires

carburizing flame contains a higher proportion of acetylene than oxygen

chaps protective covering to provide extra leg protection against heat and sparks or hot particles

check valve device that restricts movement of gas to one direction only

chipping hammer double-ended tool to chip slag in corners, edges, and ripples, and for general cleaning

concave curved like a piece of the inside of a circle

constant current when the current does not change based on workload

constant voltage when the voltage does not change based on workload

consumable insert empty box affixed to the reference line opposite a groove-weld symbol

contour symbol line drawn above the reference line that indicates the shape of the finished weld

convex curved like a piece of the outside of a circle

corrosive having a chemically destructive effect on metals

cracked opened slightly and closed immediately

cutting torch mixes a fuel gas for heating flames and a stream of pure oxygen to cut metal

cylinder used to contain fuel gas and oxygen under pressure; usually made of steel

decibels units of measurement of noise levels

direct current (DC) electric current that flows in a constant direction

direct current electrode negative (DCEN) or **straight polarity** electrode lead is the negative side and the workpiece lead is the positive side of the circuit

direct current electrode positive (DCEP) or **reverse polarity** electrode lead is the positive side and the workpiece lead is the negative side of the circuit

draglines lines that appear on the oxygen-cut surface

duty cycle amount of time a welding machine can run without overheating

electrode metal rod that is coated to provide an oxygen-free atmosphere, arc control, and a slag layer to protect the weld when cooling

electrode holder clamping device used to hold the electrode for welding

face shield protective covering for the face that can be clear for grinding or cutting

ferrous metal that contains iron

ferrous metal metal that contains iron

field weld flag that juts away from the reference line, indicating that a weld should be made at the installation point or the place of construction, not at the shop

fire blanket covering used to protect personnel and equipment

flash point amount of heat required for a substance to burst into flame

flashback arrestor safety device that prevents flame from traveling upstream of the cutting tip

friction lighter device that uses flint and a striking motion to start a spark

fuel gas gas used together with oxygen to preheat the metal to be cut; usually propane, acetylene, MPS (MAPP®), or natural gas

fumes vapor, gas, or smoke by-products of welding

groove-weld symbol show the shape of the groove; different meanings dependent on the size and shape of the joint

ground clamp produces a ground between the welding machine and the work

hot work any work involving burning, welding, or similar operations capable of initiating fires or explosions

hydrocarbon gas gas made of hydrogen and carbon only; present in petroleum and natural gas

inverter changes frequency to convert high-voltage, low-amperage AC current to low-voltage, high-amperage DC current

ionizing radiation capable of displacing electrons from atoms, creating charged atoms or ions

kerf width of the cut

manifold chamber used in conjunction with a single-stage regulator to reduce initial gas pressure before it reaches the regulator

MAPP® brand of acetylene-based fuel gas used in oxyfuel cutting

melt-through solid half-circle attached to the reference line, indicating complete joint penetration with root reinforcement in welds made from one side

methane colorless, odorless gas that makes up a large portion of natural gas

MPS cutting gas acetylene-based fuel used in oxyfuel cutting

natural gas fuel gas used in oxyfuel cutting

neutralizing flame correct flame for efficient cutting

noise unpleasant or loud sounds; can damage hearing

non-ionizing radiation visible, ultraviolet, or infrared radiation; heat or light transfer from welding work

nonferrous metal that does not contain iron

notch gouge in an oxygen-cut surface; deeper than the overall surface roughness

output slope change in relationship between the output voltage and the output curent or amperage of the machine as the current or welding workload is changed

oxidizing flame contains a higher proportion of oxygen than fuel gas

oxyfuel cutting (OFC) using oxygen and a fuel gas to cut a ferrous metal

personal protective equipment (PPE) equipment, such as goggles, helmets, gloves, and so on, specifically designed to protect exposed or vulnerable body parts when welding

plug- or **slot-weld symbol** small open box on the arrow side of the reference line

polarity direction in which current travels

propane used in oxyfuel cutting

rectifier converts AC to DC

reference line information about the welding process is located on this line; a required weld symbol

regulator device that converts gasses under high pressure to lower pressures suitable for cutting; comes in single-stage or two-stage varieties

roughness recurring peaks and valleys in the oxygen-cut surface

scratch start method of starting an arc on a scratch tab or beyond the weld starting point

shade rating assessment of how dark a lens is; measured on a numeric scale—the smaller the number, the lighter the shade

shielded metal arc welding (SMAW) or **stick welding** process that uses an electric current (AC or DC) to form an arc between an electrode coated in flux and the metals to be joined

slag combination of the heated steel, the electrode, and the electrode covering; protects the weld from the air during cooling

sleeves can be worn over existing clothing to protect the arms

soldering fusing objects with any of various alloys applied to the joint between metal without heating the objects to the melting point

spacer rectangle that sits within the reference line and indicates that joint spacer should be used

spats designed to protect the ankle area from exposure to welding particles

stinger common term for the electrode holder

stringer bead bead made with very small movements; should be no more than twice the width of the diameter of the electrode

swaged collar collar that has been crimped with a special tool to fit the hose

tail used only when there is additional information supplementing the required instructions found on the reference line

tap start method of starting an arc by tapping the tip of the electrode against the workpiece at the point where the weld is to start

top edge rounding top edge of an oxygen-cut surface that has been melted by the welder

transformer produces and regulates AC entering the machine

travel speed rate of deposition of welding bead

visor shaded or clear viewing window in a welding helmet

volatile evaporating rapidly; passing off readily in the form of vapor

weave bead bead made with small movements; should be no more than four times the width of the diameter of the electrode

weld symbols lines formed in a certain way that indicate welding actions that should take place

weld-all-around circle around the junction of the arrow and the reference line

welding helmet specially made plastic or leather shield to protect the face and eyes from welding by-products

wire brush removes rust, slag, and other dry materials from a workpiece

references & credits

The following references were used in the creation of the Cutting and Shielded Metal Arc Welding workshop.

References

American Welding Society, www.aws.org/, ANSI Z49.1 (1999)

Atlas Welding Accessories, Inc., www.atlasweld.com

Electronic Library of Construction Occupational Health and Safety, www.cdc.gov/elcosh/index.html

The Lincoln Electric Company, www.lincolnelectric.com/default.asp

Miller Welding Products, www.millerwelds.com/

U.S. Department of Labor, Occupational Safety and Health Administration, www.osha.gov/pls/oshaweb/owadisp.show_document?p_table=STANDARDS&p_id=10696

Sacks, Raymond J., and Bohnart, Edward R. *Welding: Principles and Practices*, 3rd ed. New York: McGraw-Hill Higher Education, 2005.

Welding.com, www.welding.com

Photo Credits

Front Cover

Jupiterimages Corporation/Getty Images, www.gettyimages.com

Front Matter

Decorative images used in Front Matter pages are subject to the following copyrights: Jupiterimages Corporation/Getty Images, www.gettyimages.com; Lincoln Electric, www.lincolnelectric.com; Shutterstock, www.shutterstock.com

Chapter 1

Decorative images used in this chapter are subject to the following copyrights: Jupiterimages Corporation / Getty Images, www.gettyimages.com; Lincoln Electric, www.lincolnelectric.com; Shutterstock, www.shutterstock.com

Figures 1, 2, 3, 4, 5, 7, 8, 9, 10, Lincoln Electric, www.lincolnelectric.com

Chapter 2

Decorative images used in this chapter are subject to the following copyrights: Jupiterimages Corporation/Getty Images, www.gettyimages.com; Lincoln Electric, www.lincolnelectric.com; Shutterstock, www.shutterstock.com

Chapter 3

Decorative images used in this chapter are subject to the following copyrights: Jupiterimages Corporation/Getty Images, www.gettyimages.com; Lincoln Electric, www.lincolnelectric.com; Shutterstock, www.shutterstock.com

Figure 9, Victor Thermadyne, www.thermadyne.com; Figure 10, Victor Thermadyne, www.thermadyne.com

Chapter 4

Decorative images used in this chapter are subject to the following copyrights: Jupiterimages Corporation/Getty Images, www.gettyimages.com; Lincoln Electric, www.lincolnelectric.com; Shutterstock, www.shutterstock.com

Chapter opener, Figures 2, 3, 7, Lincoln Electric, www.lincolnelectric.com; Figure 1, Miller Electric Manufacturing Company, www.millerwelds.com; Figures 8, 9, 13, Victor Thermadyne, www.thermadyne.com; Figure 12, Atlas Welding Accessories, www.atlaswelding.com; Figure 14, Shutterstock, www.shutterstock.com

Chapter 5

Decorative images used in this chapter are subject to the following copyrights: Jupiterimages Corporation/Getty Images, www.gettyimages.com; Lincoln Electric, www.lincolnelectric.com; Shutterstock, www.shutterstock.com

Figures 1, 3, 11, 12, Lincoln Electric, www.lincolnelectric.com

Chapter 6

Decorative image used in this chapter are subject toe the following copyrights: Flickr, www.flickr.com; Jupiterimages Corporation/Getty Images, www.gettyimages.com; Lincoln Electric, www.lincolnelectric.com; Shutterstock, www.shutterstock.com

Chapter opener, Lincoln Electric, www.lincolnelectric.com